Pesticides and Their Effects on Biological Organisms

Pesticides and Their Effects on Biological Organisms

Edited by:

Levitah Castil Mapatac

Delve
PUBLISHING

www.delvepublishing.com

Pesticides and Their Effects on Biological Organisms

Levitah Castil Mapatac

Delve Publishing
2010 Winston Park Drive,
2nd Floor
Oakville, ON L6H 5R7
Canada
www.delvepublishing.com
Tel: 001-289-291-7705
001-905-616-2116
Fax: 001-289-291-7601
Email: orders@arclereducation.com

ISBN: 978-1-77361-034-4 (Hardcover)

ABOUT THE EDITOR

Levitah Castil Mapatac

Levitah C. Mapatac is an Associate Professor, researcher, writer and a PhD holder from the University in Science and Technology of Southern Philippines. Her interest varies from different chemistry concepts and applications and currently doing research in traditional medicines, water waste pollution and environmental chemistry in the aspect of contamination pathways, profilling and rehabilatation of mining waste.

TABLE OF CONTENTS

PREFACE

Pesticides plays an important role in the modern technology especially in the field the pre and post harvest application in agriculture, in industry, homes and offices. The relative importance of pesticides has increased despite the availability of alternatives (organic pesticides) to highly proprietal chemical pest control such as varietal resistance and integrated pest management (IPM).Though alarmingly their is an increasing concenr on the adverse effects of pesticides to the environment, human health, and even the non-target organisms but little scientific studies had been done by the government to address this problem. This book provides vast information on the different aspects of pesticides such as its types, classification, functions, hazards, metabolism, it's toxicity and most of all its harm to human health and non-target aquatic and animal species.

The sub-classes of pesticides that includes the herbicides, insecticides, fungicides, rodenticides, pediculocides and biocides had been identified in the different chapters and what is their exact microbial activity, interactions, exact role and how it will contribute to th break down of the persistent chemical residues in a pesticide compounds.Many pesticides can be grouped into chemical families which in return this pests can develop resistance to the which is sometimes dose dependent which can indiscriminate use of pesticide, necessitating a new pesticide. Alternatively an increased dose of the pesticide can be used to counteract the resistance, which will worsen environment further. Prominent insecticide families include organochlorines, organophosphates,carbamates, and endosulfan in natural environment.

Content Coverage

Chapters One and Two are intended to present an overview of what are the classification of pesticides- microbial, insecticides, nematicides, fungicides, molluscicides and herbicides and its biological template, properties of this types of pesticides.

Chapters Three and Four present the effects pesticides to human health and to the environment especially those containing the highly resistant kind of chemical compound of organochlorines, organophosphates and carbamates. Common symptoms and manifestations of the pesticide poisoning due to the presence of OP, organochlorine derivativeswhich will be harmful to the consuming public

Chapters Five and Six include information about the health problems caused by the magnification of pesticide residues and the treatment of common pesticide poisoning. This deals with the environmental concerns on the aspects of biodiversity, wildlife, aquatic environment and its systems.

Chapters Seven and Eight focus on the properties of pesticide in different category, functionalities and impact in the ecosystem. In this particular chapters it deals with the extent of exposure and different preventive measures to counter the harmful effects of pesticide poisoning on human health and the environment.

Chapters Nine and Ten cover the ecological impact of organochlorines and its derivatives- endosulfan. This also touches the occurence of biomagnification and how microbial degradation can help in the persistence of this chemical compound.

Chapters Eleven to Thirteen emphasizes the different properties and functions of the sub-types of pesticides:microbial, insecticides, nematicides, fungicides, molluscicides and herbicides. Its mechanisms, behavior, formulation, dissipation and most of all its toxicity level in terms of lethal dose (LD_{50}) and lethal concentrations (LD_{50}).

Editor

1
CHAPTER

PESTICIDE IN THE ENVIRONMENT

INTRODUCTION

Pesticides are synthetic products which can have negative effects on the environment and are dangerous chemicals that are applied to different living organisms such as plants and soil which has residual effects. While a pesticide is meant to eradicate a specific pest, a very large percentage of pesticides reach a destination other than their target. They tend to enter in earth's ecosystem, especially the air, water, soil, and finally into the food chain which includes plants, animals and human beings through the food chain. Pesticides are chemical compounds needed to eradicate nuisance pests and other organisms which will hinder man's activities in producing food products from plants.

Examples of pesticides include fungicides, rodenticides, herbicides, insecticides, nematicides, molluscides and other plant growth hormones. Organochlorine (OC) , the main component present in the insecticides used in eliminating selected diseases cause by pests such as malaria and typhus, was restricted in 1960's in some developed countries. Another synthetic insecticide during the 1960's known as organophosphate (OP) insecticides, in 1970's carbamates and pyrethroids in 1980's was successfully being use as herbicides and fungicides in controlling pest for agricultural production.

Practically pesticide must be dangerous to the targeted pests, but not to non-target species especially to human being's. But some instances wherein the abuse used by pesticide plays havoc in human beings life and even to aquatic and forms of life.

WHAT IS THE ENVIRONMENT?

A part of the ecosystem consisting of plant, animal and other organisms which can be affected by addition of synthetic chemicals such as as pesticide which can affect the soil, water and other life's vegitation. Effect could either contaminate its natural system by poisoning the target and non-target organisms.

What is a pesticide?

The environmental effect of pesticide awareness came about when renowned environmentalist Rachel Carson (1962), the author of the Silent Spring, denounced the heavy use of pesticides in the environment. This became an eye opener on the side effects of the use of this insecticide in control of malaria infestation due to mosquito bites.

Figure 1.1: Rachel Carson played a key role in birth of environmentalism in the 20th century (1907-1964)

Actually DDT, chemically known as *-dichloro diphenyl trichloroethane*, becomes popular during World War and was the first organic pesticide discovered and even known as the " wonder chemical" as solution for different pest control. DDT fared for about forty years until it's ill effect was exposed by Rachel Carson due to its long lasting effect in the environment such as:

- ***Food supplies:*** USDA found DDT breakdown products in 60% of heavy cream samples, 42% of kale greens, 28% of carrots and lower percentages of many other foods.
- ***Body burden:*** DDT breakdown products were found in the blood of 99% of the people tested by CDC.
- ***Health impacts:*** Girls exposed to DDT before puberty are 5 times more likely to develop breast cancer in middle age, according to the President's Cancer Panel.

Synthetic organic compounds used as pesticides release toxic substances intentionally into our environment to kill living things and even persist as chemical residue in the soil. Example of these substances are weed killers (herbicides), insects (insecticides), fungus (fungicides), rodents (rodenticides), and others as shown in figure 1.2.

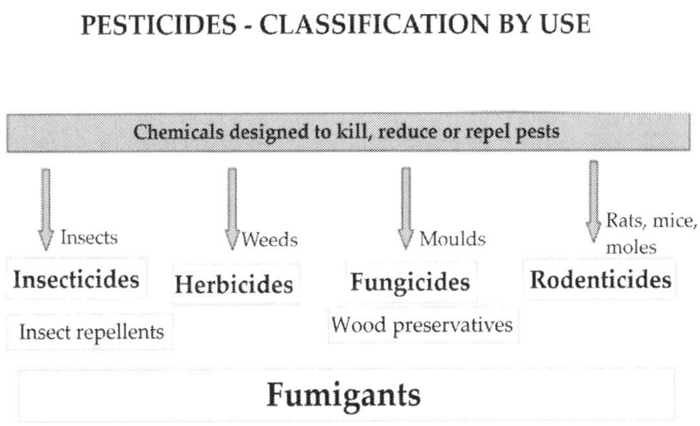

Figure 1.2: Classification of Pesticides with its Uses

Further, the use of these toxic pesticides had become a common practice among the people who wants to get rid of pest management around the world. Pesticides are used almost everywhere for domestic use, in the field of agriculture for the increase of yield, in the homes, schools, offices and other aspects that needs to eradicate the presence of pests.Thus, pesticide can be found in the air that we breathe, the food the we eat and most of in the water that we drink, or its everywhere.The table below shows the different pesticides with its uses.

Table 1.1: Types of Pesticides according to National Pesticide Information Center in coordination with US EPA.

Types of Pesticides	Uses of Pesticides
Algaecides	are used for killing and/or slowing the growth of algae.
Antimicrobials	control germs and microbes such as bacteria and viruses.
Biopesticides	are made of living things, come from living things, or they are found in nature.
Dessicants	are used to dry up living plant tissues.
Defoliants	cause plants to drop their leaves.
Disinfectants	control germs and microbes such as bacteria and viruses.
Fungicides	are used to control fungal problems like molds, mildew, and rust.
Herbicides	kill or inhibit the growth of unwanted plants, aka weeds.
Illegal and Counterfeit Pesticides	are imported or sold illegally.
Insecticides	are used to control insects.
Insect Growth Regulators	disrupt the growth and reproduction of insects.
Minimum Rish Pesticides	are exempt from EPA registration, but many states require them to be registered.
Miticides	control mites that feed on plants and animals. Mites are not insects, exactly.
Molluscicides	are designed to control slugs, snails and other molluscs.
Mothballs	are insecticides used to kill fabric pests by fumigation in sealed containers.
Natural and Biological Pesticides	control pests using things found in nature, or man-made versions of things found in nature.
Ovicides	are used to control eggs of insects and mites.
Pheromones	are biologically active chemicals used to attract insects or disrupt their mating behavior. The ratio of chemicals in the mixture is often species-specific.
Plant Growth Regulators	are used to alter the growth of plants. For example, they may induce or delay flowering.

Repellents	are designed to repel unwanted pests. often by taste or smell.
Rodenticides	are used to kills rodents like mice, rats, and gophers.
Synergists	make certain pesticides more effective, but they are not effective when used alone.
Wood Preservatives	are used to make wood resistant to insects, fungus and other pests.

USE OF PESTICIDES DURING ANCIENT TIMES

The origin of pesticides was recorded as the early part of 2500BC; pesticides were accidentaly discovered when the Sumerians applied sulfur to their bodies to drive away the mites as well as the ancient Egyptians used pesticides as recorded in the oldest medical documents in an Eber papyrus paper dated back to 1550BC, wherein Egyptians had made about 800 different ways of poisons and pest repellents.

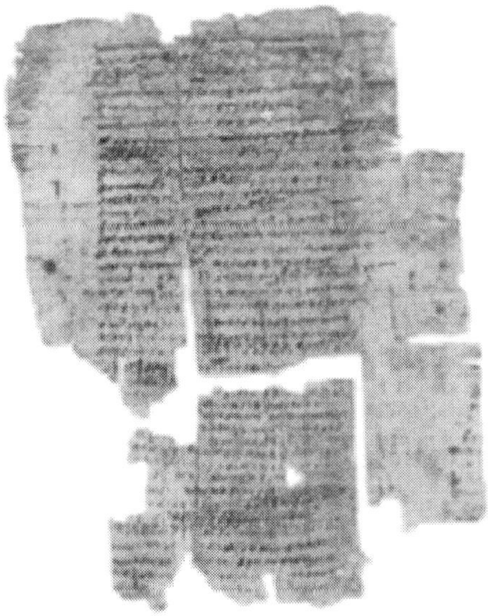

Figure 1.2: The Ebers Papyrus written by Egyptian alchemists in 1550 BC.

In Greece around 1000 BC, Homer described how Odysseus "fumigated the hall, house and court with burning sulfur to control pests," and this was followed by the Chinese who were using compounds made from mercury and arsenic to control body lice. Even in the biblical setting from the Old Testament (First Kings 8:37) it lists blights and mildews as among the great scourges of mankind. In approximately 300 B.C. Theophrastus, the father of modern botany was the first to document the application of olive oil in the roots of an affected trees attacked by different pests. He also noted that certain weeds were associated with specific crops.

During 234 BC to 149 BC- Cato, the Roman Censor applied oils in spraying hell bore plant to eradicate rodents and insects.

The major outbreak started in application especially during occurence the destroy of agricultural crops in Ireland during the 19th century as potato blight (*Phytopthora infestans*) and this was followed by the application of synthetic compounds as pesticides application with the presence of chemical such as arsenic, pyrethrum, lime sulfur, and mercuric chloride was used.

The period of World War II, the application of synthetic compounds used such as the Paris green, arsenic compounds of lead and calcium, selenium compounds, lime–sulfur, pyrethrum, thiram, mercury, copper sulfate, derris, and nicotine were used and many more to increase cropping yields. After World War II the use of pesticides mushroomed, and there are currently more than 1,600 pesticides available and about 4.4 million tons used annually, at a cost of more than $20 billion and the pesticides consumption was greater during 1995 to 2000 and lowers its used in years 2000 to 2005 as shown in table 1.2 .United States accounts for more than 25 percent of this market and herbicides accounts almost half of the pesticide used in agricultural sectors (figure 1.3).

Table 1.2: Consumption of pesticides (group wise) in MT(metric tons) during 1995-1996 to 2004-2005 (source:http://ncipm.org.in/asps/pesticide)

Pesticide Group	1995-96	1996-97	1997-98	1998-99	1999-2000	2000-01	2001-02	2002-03	2003-04	2004-05
Insecticide	38,788	34,665	33,379	30,469	28,926	26,756	29,839	28,897	25,627	25,929
Fungicide	10,563	9,969	10,054	10,428	8,435	8,307	9,222	10,712	9,087	6,397.4
Herbicide	6,040	7,060	7,103	7,292	7,369	7,299	6,979	7,857	5,610	7,364
Others	5,869	4,420	1,703	968	1,465	1,222	1,308	1,398	438	1,660
Total	61,260	56,239	52,239	49,157	46,195	43,584	47,348	48,146	40,762	41,350.4

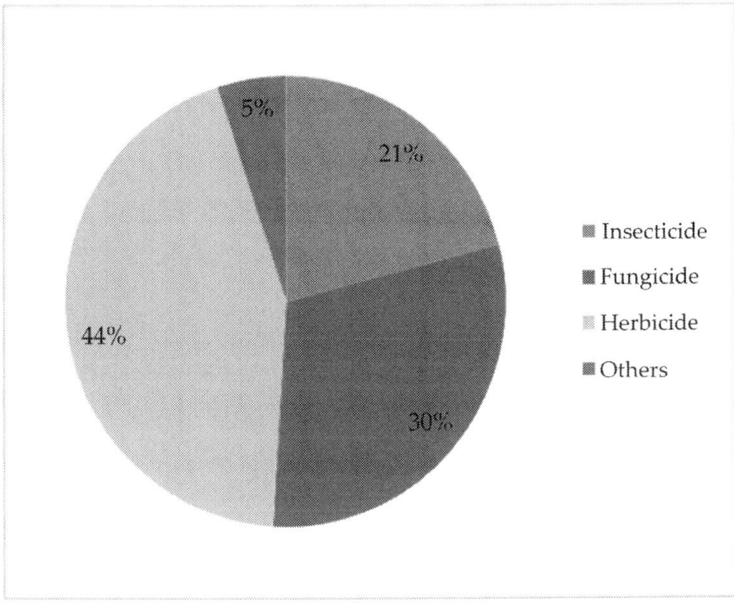

Figure 1.3: Pesticide Use Pattern in the World (source:http://www.cibrc.nic.in/reg_products.htm)

TYPES OF PESTICIDES

Older types of insecticides

Organochlorine is an insecticide, commonly known to contain DDT. DDT is chemically named as dichlorodiphenyl-trichloroethane and was first discovered in Switzerland in 1939. This was commonly applied for human diseases such as body lice and other disease vectors, especially for agricultural pests. Another type of chloronated insecticides are the benzene hexachloride, which were discovered during World War II, followed by another organochlorine containing insecticide toxaphene or the heptaclor.

Another organochlorines with aromatic compounds was discovered in the form of aldrin and diedrin and followed by endrin, endosulfan and isobenzan. Then, during 1940's and 1960's extensive chlorinated hydrocarbons were developed such methoxychlor, dieldrin, chlordane, toxaphene, mirex, kepone, lindane, and benzene hexachloride. As neurotoxicants, United States had given a law in stopping the of organochlorine pesticides unless registered for used.

Organophosphate as an insecticides started in Germany and was purposely used as nerve gas during World War II but was this compound was fully used as insecticides and broaden up as tetraethyl pyrophosphate (TEPP) and parathion, demeton, methyl schradan, phorate, diazinon, disulfoton, dimethoate, trichlorophon, and mevinphos have been registered.

The acetylcholine (Ach) acts as the neurotransmitter of the inhibiting enzyme cholinesterase (Che) that blocks the activity and causes paralysis and even death to insects and sometimes mammals. The good effect of organophospate has an environmental significance because this has no chemical residue and does not bioaccumulate in animals.

Carbaryl insecticides, with chemical name of 1-napthyl methylcarbamate, acts on nervous transmissions in insects and also through effects on cholinesterase by blocking acetylcholinereceptor. This was discovered by Union Carbide in 1958 and very popular due the incident with the Bhopal disaster in 1984.This insecticide was synthesized through treating methyl isocyanate with 1-napthol. Carbryl is an example of an insectice with broad-spectrum and has moderate toxicity, low persistence, and rarely bioaccumulates in the environment.

Botanical insecticides are plant extracts and minerals that have insecticidal properties which are toxic to insects. Plant extracts have naturally occuring substances that act as toxins to insects called "natural insecticides" and most organic gardeners prefer to choose these botanical insecticides. Some example of these naturally occuring insecticides come from plant extracts (tobacco leaves, neem leaves, weed extract, sabadilla extracts and citrus peels which have low levels of toxicity for animals).

Table 1.3: Example of botanical insecticide with its levels of toxicity (Source: Environmental Protection Agency (EPA) and used in accordance with the provisions of the Federal Insecticide, Fungicide and Rodenticide Act (FIFRA).

Generic Name	Oral LD50	Dermal LD50	Signal Word
Pyrethrins	1,200-1,500	>1,800	Caution
Rotenone	60-1,500*	940-3,000	Caution

Sabadilla	4,000	—	Caution
Ryania	750-1,200	4,000	Caution
Nicotine	50-60	50	Danger
d-Limonene	>5,000	—	Caution
Linalool	2,440-3,180	3,578-8,374	Caution

* Toxicity varies greatly depending on type of solvent used as carrier.

Newer Insecticides

Pyrethroids such as permethrin, cypermethrin, fenvalerate and cyfluthrinare are all examples of pyrethroids. The first pyrethroids were synthesized and the structure of insecticidal pyrethrins was found with the same chemical compound found in the seeds of certain chrysanthemums.

Pyrethroids last longer than their natural counterparts and seem to have a similar mode of action interacting the regular transmission of the nerve impulses which is being channeled by the nerve cells with the aid of inorganic elements of potassium and sodium.

Insect Growth Regulators (IGRs), insects undergo through a process known as growth development such as shedding of one's skin and molting wherein a hormone is needed for this life changing process for the insects. Thus, insect growth regulators such as methopreneandpyriproxifen will hamper the change in form of the insect by IGRs will copy its action and results in the disruption of the physiological process of the insects. The insects will result in immature and ready for the change into adulthoood or pupa stage, finally causes the insects to die but has safety for people.

Microbial Insecticides

Microorganismsnow a days are made used to attack insects specifically viruses that takes over the function of the insects. In most cases the viruses that function as insecticides will be eaten by insects. This results by making copies of themselves, the insect's cells will burst and die due to this action.

Insecticides using a bacterium *Bacillus thuringiesis* works wherein the insects eats the bacteria. When activated in the insect gut, they destroy the selective permeability of the gut wall. For instance, in application to flies and beetles, the first strains were toxic only to *Lepidoptera,* and the

Bacillus thuringiesis can incorporated to the insects body which will be deadly to the insects.

Thus, allows the bacteria to take over the insect's body especially the gut lining and invading the insects body and finally dies.Whille fungal insecticides can attack insects from the outside. By making the spores land on the insect's body and grow right into it,eventually penetrating inside and growing throughout the body.

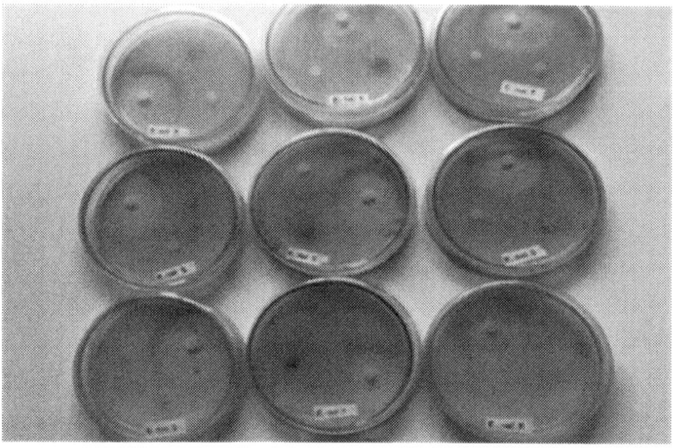

Figure 1.4: Application of Pesticide in Microbial Analysis in the effectivity of bacteria in plant extract as insectide

Other Insecticides

***Fipronil*is** a phenylpryazol insecticide used to control some agricultural crops attack by pests and parasites . Its action is to to block the nerve cells and disrupts the normal nerve function. Fipronil also is being used to treat rice seeds due to the presence of ticks and lies. Avermectins have this same mode of action.

Soaps are used as "non-chemical" insect controls and composed mainly of potassium salts of fatty acids. When soaps contact the outer body covering of insects, they penetrate and dissolveinto the cells causing the cell fluids to leak out and resulting to to collapse and dehydration of the cells and finally causes death to the insects.

Oils sprayed on mite and aphid eggs are designed to kill the eggs by smothering them, preventing exchange ofoxygen.

Nematicides

Soil nematicides are chemical pesticide which is used to kill plant tiny parasitic worms that lives in water or soil. Nematicides are toxic substances that has a broad spectrum which is highly volatile and use as soil fumigants by spraying to the roots of the affected plant.

Example of this nematicides are the dichlopropene, methyl isocyanate, chloropicrin, and methyl bromide.Other which could only be applied through contact are the aldicarb, dazomet, and metham sodium.

This nematicides has high toxicities range which mamalian and other organisms coming from the plant and animal group could be affected with the application of this pesticide. Although **transient** in soil, they may have drastic ecological effects on soil systems.

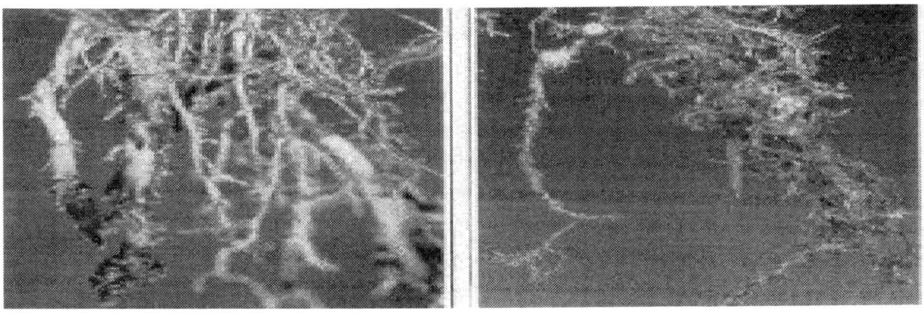

Figure 1.5: Plant roots being affected by parasitic worms.

Molluscicides

Molluscicides are pesticides which kills mollusks such as metaldehyde and methiocarb that baits against slugs and snails. Although of high mammalian toxicity, they cause few problems that cause accidental death to wild mammals. Several molluscicides, used to control aquatic snails, *N* -trityl morpholine, copper sulfate, niclosamine, and sodium pentachlorophenate, are toxic to fish.

Figure 1.6: Golden Apple Snail applied with plant extract as molluscicide.

Herbicides

These are chemicals used to destroy unwanted plants such as weeds and was discovered during the 1940's.Two types herbicides are the inorganic compounds such as copper sulfate, sodium chlorate, and sodium arsenite while organic compounds such as chlorophenoxy ,dinitrophenols, bipyridyl,carbamates and amide herbicides. Historically, inorganic compounds were the first used and made into commercial products. There has been over a long period a continuous effort to develop herbicide compounds that are more selective—that affect weeds, as opposed to desirable plants.

Figure 1.7: Application of herbicides in rice production.

Fungicides

Fungicides are pesticide that is used to control the infestation of fungus causing diseases in plants.This has low toxicity with the mamalian and wide spectrum differing chemical structures and have a relatively narrow spectrum of toxicity to soil-inhabiting and aquatic organisms. Their greatest environmental impact is toxicity to soil microorganisms, but these effects are short term.

Figure 1.8: Fungus infestation in corn cob plant.

Fungicides are either chemicals or biological agents that inhibit the growth of fungi or fungal spores. Modern fungicides do not kill fungi, they simply inhibit growth for a period of days or weeks. Fungi can cause serious damage in agriculture, resulting in critical losses of yield, quality and profit.

Fungicides are used both in agriculture and to fight fungal infections in animals. Chemicals used to control oomycetes, which are not fungi, are also referred to as fungicides as oomycetes use the same mechanisms as fungi to infect plants (Latijnhouwers et al., 2000).

Fungicides can either be contact, translaminar or systemic. Contact fungicides are not taken up into the plant tissue, & only protect the plant where the spray is deposited; translaminar fungicides redistribute the fungicide from the upper, sprayed leaf surface to the lower, unsprayed surface; systemic fungicides are taken up & redistributed through the xylem vessels to the upper parts of the plant. New leaf growth is protected for a short period.

Most fungicides that can be bought retail are sold in a liquid form. The most common active ingredient is sulfur, present at 0.08% in weakerconcentrates, and as high as 0.5% for more potent fungicides. Fungicides in powdered form are usually around 90% sulfur and are very toxic. Other active ingredients in fungicides include neem oil, rosemary oil, jojoba oil, and the bacterium Bacillus subtilis .

Fungicide residues have been found on food for human consumption, mostly from postharvest treatments (Brooks & Roberts, 1999). Some fungicides are dangerous to human health, such as vinclozolin, which has now been removed from use (Hrelia, 1996), FCX and DFB thatare used as pesticides to control pests and they have many side effects on natural non-target organisms (Rouabhi et al., 2009). In this chapter, we will develop the fungicides and their toxicity on biological and ecological systems.

REFERENCES

1. Bohmart, B.L. (1997). *The Standard Pesticide Users Guide,* 4th edition. London: Prentice-Hall International.
2. Carson, Rachel. (1963). *Silent Spring.* London: Hamish Hamilton.
3. Ekstrom, C., ed. (1994). *World Directory of Pesticide Control Organizations.* Farnham, U.K.: British Crop Protection Council.
4. Leng, M.L.; Leovey, E.M.K.; and Zubkoff, P.L., eds. (1995). *Agrochemical Environmental Fate: State of the Art.* Boca Raton, FL: CRC Press.
5. Pimentel, D., Lehman, H., eds. (1993). *The Pesticide Question: Environment, Economics, and Ethics.* New York: Chapman and Hall.
6. Rand, G.M., ed. (1995). *Fundamentals of Aquatic Toxicology: Effects, Environmental Fate and Risk Assessment.* Washington, D.C.: Taylor and Francis.
7. Smith, R.P. (1992). *A Primer of Environmental Toxicology.* Philadelphia: Lea and Febiger.
8. Ware, G.W. (1994). *The Pesticide Book,* 4th edition. Fresno, CA: Thomson Publications.

9. Impetus for sowing and the beginning of agriculture: Ground collecting of wild cereals; M.E. Kislev, E. Weiss and A. Hartmann, Proceedings of the National Academy of Sciences, 101 (9) 2692-2694 (2004)
10. http://www.weizmann.ac.il/kimmel-arch/pdf/17_Kislev2004.pdf
11. Primal Seeds, Origin of Agriculture http://www.primalseeds.org/agricult.htm Economic Benefits of Pest Management; R. Peshin, Encyclopedia of Pest Management, pages 224-227, Pub. Marcel Dekker, 2002
12. The History of Pesticides, Organic Pesticides, September 19th 2008 http://blog.ecosmart.com/index.php/2008/09 19/the-history-of-pesticides/
13. 14 . History of Horticulture, Roman Agricultural History; J. Janek, Purdue Universityhttp://www.hort.purdue.edu/newcrop/Hort_306/text/lec18.pdf
14. Forerunners of Pesticides in Classical Greece and Rome; A.E. Smith and D.M. Secoy, J. Ag. Food Chem. 23 (6) 1050 (1975) Seehttp://www.hort.purdue.edu/newcrop/history/lecture18/r_18-1.html
15. Pyrethrum, The Natural Insecticide; Equatorial Health Services http://www.equatorialhealth. com/pyrethrum.html
16. A Compendium of Inorganic Substances Used in European Pest Control before 1850; A.E. Smith and D.M. .Secoy, J. Ag. Food Chem. 24 (6) 1180 (1976) See http://www.hort.purdue.edu/newcrop/history/lecture31/r_31-1.html
17. A History of Crop Protection and Pest Control in our Society; CropLife Canada (2002) http://www.croplife.ca/english/pdf/Analyzing2003/T1History.pdf
18. Pesticide Usage in the United States: History, Benefits, Risks, and Trends; Bulletin 1121, November 2000, K.S. Delaplane, Cooperative Extension Service, The University of Georgia College of Agricultural and Environmental Sciences http://pubs.caes.uga.edu/caespubs/pubs/PDF/B1121.pdf
19. Wessels Living History Farm, York, Nebraska; Farming in the 1950s & 60s http://www.living historyfarm. org/farminginthe50s/pests_08.html

20. Wessels Living History Farm, York, Nebraska; Farming in the 1930s http://www.livinghistoryfarm.org/farminginthe30s/pests_04.html
21. Silent Spring, 40th Anniversary Edition, Rachel Carson, Houghton Mifflin Harcourt, 2002 ISBN
22. New Insecticide Modes of Action: Whence Selectivity? J. Coats, Iowa State University, Ames, Iowa, USA
23. http://www.slideworld.org/viewslides.aspx/New-Insecticide-Modes-of-Action-Whence-Selectivity-ppt-42841
24. A Short History of Fungicides, V. Morton and T. Staub, APSnet, March 2008 http://www.apsnet.org/online/feature/fungi/
25. OECD SERIES ON PESTICIDES, Number 8, Report of the OECD/FAO Workshop on Integrated Pest Management and Pesticide Risk Reduction, April 1999 http://www.olis.oecd.org/olis/1999doc.nsf/LinkTo/NT00000FBE/$FILE/04E94320.PDF
26. Centers for Disease Control and Prevention, Fourth National Report on Human Exposure to Environmental Chemicals –http://www.cdc.gov/exposurereport/pdf/FourthReport.pdf
27. U.S. E.P.A. Consumer Factsheet on Lindane -http://www.epa.gov/safewater/dwh/c-soc/lindane.htmlATSDR ToxFAQs –Chlordane -http://www.atsdr.cdc.gov/tfacts31.html
28. Handbook of Agriculture, ICAR, 2007
29. http://cei.org/node/20957
30. http://oregonstate.edu/~muirp/pesthist.htm
31. http://www.alanwood.net
32. http://www.btny.purdue.edu/Pubs/PPP/PPP-31.pdf
33. http://www.cibrc.nic.in/reg_products.htm(2007)
34. http://www.ncipm.org.in/asps/pesticides

2
CHAPTER

BIOLOGICAL EFFECTS OF PESTICIDES

INTRODUCTION

The main objective on the application of pesticide is to help farmers yield a better harvest by eradicating pests and other forms of life that hampers or kill some forms of life. US EPA defines pesticide as follows: "A pesticide are synthetic chemical compounds whose sole purpose is preventing, destroying, repelling, or mitigating any pest. Sometimes this term is misunderstood as pertaining only to insecticides, though the term pesticide could mean also to the application of herbicides, fungicides, and various other substances used to control pests."

Pesticide main components contains both "active" and "inert" compounds. Active compounds are what kill the pest, and inert ingredients help the active ingredients to work more effectively.

Types of pesticides

There are two large classes of pesticides the first one is insecticides (insect killers) and herbicides known as weed killers. Other smaller classes of pesticides are fungicides (fungi and molds), rodenticides (for mammals especially rats) and antimicrobials (application to microorganisms specially bacteria and viruses).

Antimicrobial pesticides has a wider application especially in food preservation, medical sterilizers, disinfecting homes, schools, commercial buildings and any institution or offices.

Figure 2.1: Different Commercial Pesticides available in the market.

Functions of pesticides

Pesticides works by controlling some essential functions in different aspects in pest specifically its physical, chemical and biological metabolic behavior. Pesticides normally are toxic to target pests once applied and may harm other organisms which is not the main target of the application. Though some pesticides are not toxic to non-target such as insect repellants or attractants (such as mosquitoes and pests repellents) sterilizing substances (controls the reproduction and fertilization of some pest), some type of defoliants and dessicants(sprayed on plants to make their leaves fall off) and pesticide that enhances the action of another pesticide without being affected(example are piperonly butoxide and pyrethrum).

Furthermore, pesticides are effective in eradicating nuisance pests but because all living organisms has almost the same biological system wherein pesticides may kill or harm other organisms. Human beings who are also exposed to pesticides by using it at homes, eating foods which was applied with pesticide (chemical residues) or people who directly spraying pesticides through contaminated air (farmers who applied pesticide for pest control).

Pesticide producers should be careful enough to come up with an ideal pesticide which is highly specific only to the target organism but harmless to non-target organisms and no chemical residues should be produce specially in soil.

BIOLOGICAL PROPERTIES OF INSECTICIDE

Most chemical insecticides act by poisoning the nervous system. The main target is the center of the peripheral point of the nervous system among insects which works the same with that of mammals. A minute quantity of the application of this pesticide can be fatal to an insect, primarily because of the insect's small size and high rate of metabolism. Though at certain level this will not be harmful and fatal to human beings due to lower concentration but at the same time it may also cause harm.

The pattern of nervous system structure make it nearly impossible to design insecticides that affect only target insect pests; consequently, insecticides may affect non-pest insects, people, wildlife, and pets. Insecticide application could also harm the aquatic life in some ways hence pesticide application tend to leach out in the water ways for example of insectice carbryl (a carbamate insecticide) which was listed as one of cause of carcinogen substances which could be a source of hormone disruptor as stated by USEPA.

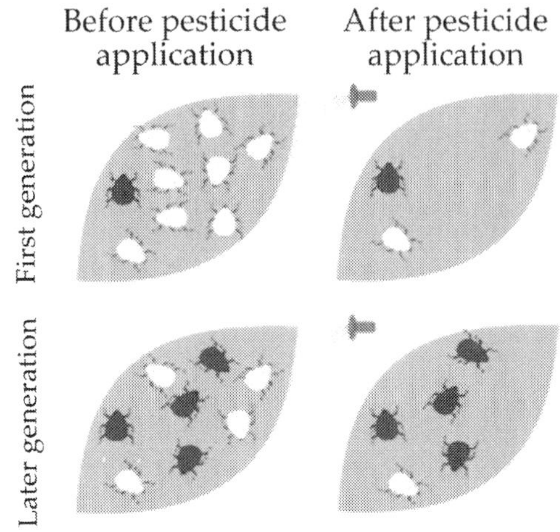

Figure 2.2: The biological pesticides before and after application.

The newly discovered insecticides of the new mellinium are designed to be more specific and less chemical residues in the environment. The most prominent classes of insecticides are organochlorines, organophosphates, carbamates, and pyrethroidsas shown in the figure 2.2 is the application of biological pesticides.

Organochlorines

Figure 2.3: An insecticide pest killer that contains DDT.

Organochlorine chemical structure differs in a way that this contains chlorine compounds that replaces most of the bigger composition of chlorinated hydrocarbons. DDT, contains organochlorines which is one of the beneficial component as insecticidal use but has dangerous risks in health and environmental factors.

Organochlorines is advantageous in a way this has cheaper manufacturing cost and more specific in the target specie and has less environmental effect especially in the soil.Organochlorines tend to have a metabolic responses such this will stop the movement of ions in compounds containing calcium, chloride, sodium, and potassium into and out of nerve cells. Depending on the specific structure of the organochlorine chemical, it may also affect the nervous system in some ways. At one time organochlorines were thought to be ideal because they are very stable, slow to disintegrates in the environment, dissolves in fats (and are therefore readily taken up by insects), and seemingly harmless to mammals. Sadly after few years its effect on the environment becomes controversial due it's persistence and its undesirable fat solubility thus's accordingly organochlorines could passed up in the food and could have a large fat accumulation to large animals and humans specifically it has undesirable effect to nursing animals and human beings.

Globally the use of organochlorines has contaminated wildlife around the world even the lower part of the world specially the Artic and Antartica regions due to excessive used and application of pesticides. DDT contaminations affects the wildlife which was exposed to pesticide and may caused the thinning of the bird's eggsheels and causes death to developing birdlings.

Organochlorines like DDT presently banned in it's used especially in the highly industrialized countries but still of used to some developing countires .Organochlorine insecticides provide many important aspects interms of it's about the desirable and undesirable properties of pesticides. Thus, organochlorine may have advantageous use but it could have undesirable substances found as pesticides.

Organophosphates and Carbamates

Organophosphates and carbamates differs in its chemical structure but this compound shows almost the same mechanism and action as an insecticides. Wherein organophosphates was first invented in the 1940's used as effective and hazardous biological warfare agents especially as a nerve gas. Newly discovered derivatives of organophosphates namely sarin, soman and VX were banned in some countries ready for disposal but still has difficulties in looking for a proper way of disposing this materials. Researchers have found out new derivatives of organophospates as insecticidal effect which would be less toxic and cost effective and specific in its reaction, less harming the non-target.

When the organophosphate parathion was first used as a substitute compound with DDT, it was think as a better solution and hope that this will be less harnful to non-target but unfortunately this leads to destruction hence this resulted to more human and animal deaths and has high toxicity more than what is expected with DDT application.

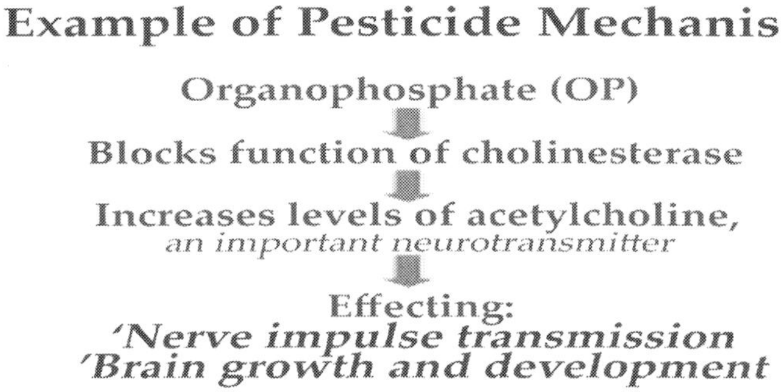

Figure 2.4: The Biological pathway on the effect of organophosphate.

Organophosphates and carbamates has an action towards the nervous system as a neurotransmitter both in insects and mammals (as shown in figure 2.4). Acetycholine, as a neutrotransmitters brings about every nerve cells to enable somebody to communicate with each other.

The action of acetycholine acts an initiator for every communication in the nerve by transmitting this to another nerve cells by stimulating every nerve ending in the cells.Thus, to stop every communication the acetylcholine as an enzyme, acetylcholinesterase breaks down the acetylcholine. Actually the organophosphates and carbamates stops the enzyme by distrupting the the functioning to the different nerve cells. As an insecticides the enzyme acetylcholinesterase acts an inhibitors.

Chemical structure differs between organophosphates and carbamates affects the action of acetylcholinesterase as an inhibitor and affects its level of efficiency. As a nerve gas, acetycholine are highly efficient as nerve blockers and permanently block acetylcholinesterase, while the commonly used pesticides block acetylcholinesterase only temporarily. The lethality of these pesticides has significant health dangers wherein researchers continued to investigate new insecticides with less environmental and health hazards effects.

Pyrethroids

Pyrethroids, a new class of insecticide synthetic in nature which was based on a natural bioactive compounds from chrysanthemum flowers. During 1980's a synthetic pyrethroids was investigated and was commercially available upon it's effective application and continued to be popular for the last 20 years.

The chemical template of pyrethroids differs from that of organochlorines, organophosphates, and carbamates though it's primary site of action is in the nervous system. Pyrethroids affects the movement of sodium ions (Na+) into and out of nerve endings in the cells, causing the nerve cells to becomeseasily attracts the neurotransmitters. Structural components between various pyrethroids can change their lethality depending on the insects and animals.Synthetic pyrethroids are more persistent in the environment than natural pyrethrum, which is unstable in light and breaks down very quickly in sunlight.

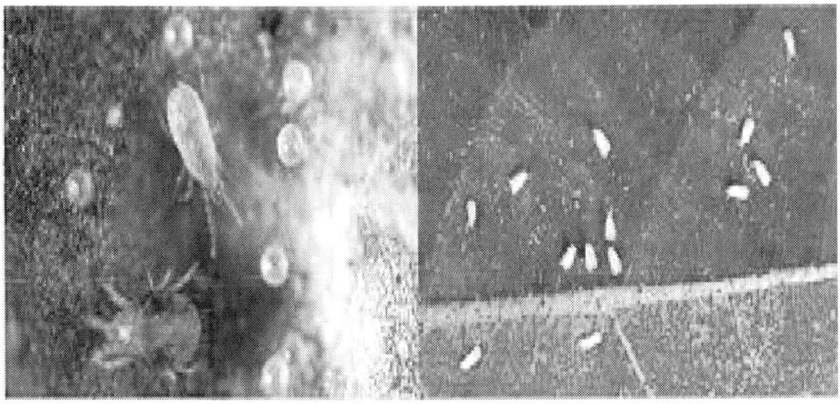

Figure 2.5: Application of phyrethroids can disrupted the natural enemies or pests.

BIOLOGICAL PROPERTIES OF HERBICIDES

Herbicides are known as a weed killers that used to damage plants and are the most rapidly growing type of pesticide. During the early part of 1930's, herbicides were nonspecific and tend to be dangerous to human beings and animals. Researchers tend to look for substitute chemicals that selectively killed plants while developing new insecticides. These chemicals are now widely used to increase food production by killing weeds that choke out or compete with food crops.

Chlorophenoxy, a well-known herbicide containing compounds that include 2,4-D and 2,4,5-T. Agent Orange a herbicide known in the 1960's, which was broadly applied in agriculture specially to kill broadleaf plants a weed that competes with the nutrients of the plants especially rice production.This was also extensively used as a chemical warfare agent to kill unwanted vegetation in an open field. Though its action mechanism of this herbicides is not understood well but the herbicides really effective as plant growth hormones.

A nonselective herbicide known as paraquat and it's related chemical compound of diquat are also lethal to mammals. Occupational or accidental exposure to paraquat can occur by ingestion, skin exposure, or inhalation, all of which can cause serious illness or death. Paraquat was sometimes used in industrialized countries but still widely used in developing countries. At one point in time it was used in marijuana plant eradication programs, but it was discontinued when a number of fatalities were observed in smokers of paraquat-contaminated marijuana.

Many other herbicides are widely used as alachlor, glyphosate, and atrazine and they have a range of actions on plants and animals.

Herbicides becomes an integral part in agriculture specially this becomes a need for economic reasons to have a good production and high yield of harvest.In connection with this is a serious limitation of many herbicides affects the non-target organisms in other words, herbicides can damage the crops of interest. The manufacturers of herbicides are working to address this problem and are increasingly turning to a new technology that would create genetically modified crops that are herbicide resistant.

Figure 2.6: Application weeds sprayed with herbicide.

An example for this is the application of herbicide Monsanto that produces the glyphosate-based herbicide RoundUp. The company also manufactures a genetically modified soybean that is resistant to RoundUp. This allows farmers to use RoundUp herbicide with the RoundUp herbicide that can be applied to soybean plants and not have to worry about killing the soybean plants. The GMO applied RoundUp applied to soybean is now widely planted, though the practice has generated considerable controversy internationally.

Biological Uptake of Herbicides

Herbicides must move from outside the plant, the leaf surface or stem, if foliar-applied, or from the soil solution, if soil- applied, to the target site inside plant cells. The driving force for most herbicide movement is diffusion.

The herbicide pathway is composed of many different media that emulate aqueous solutions or organic solvents (Figure 2.7). The cuticle on leaves and stem is waxy, therefore, water loving (hydrophilic) compounds have difficulty dissolving through this layer. Water hating oil soluble (hydrophobic) compounds may dissolve into the cuticle and never come out.

Next, the herbicide encounter a water filled area (apoplast) that bathes the cells, and the situation is reversed. Hydrophilic compounds can go into solution easily, while hydrophobic is reversed. Hydrophilic compounds can go into this solution easily, while hydrophobic materials enter and move through slowly. Additionally, water moves through this area, making its way to the edge of the leaves where most is lost to transpiration. Therefore, hydrophilic compounds can get swept along to the edge of the leaves.

Surrounding the cell is a cell memberane. This membrane is primarily made of oil-like substances called phospholipids and some proteins. Hydrophobic herbicices enter the oily membrane, hydrophobic herbicides exit slowly, while hydrophilic herbicides exit rapidly.

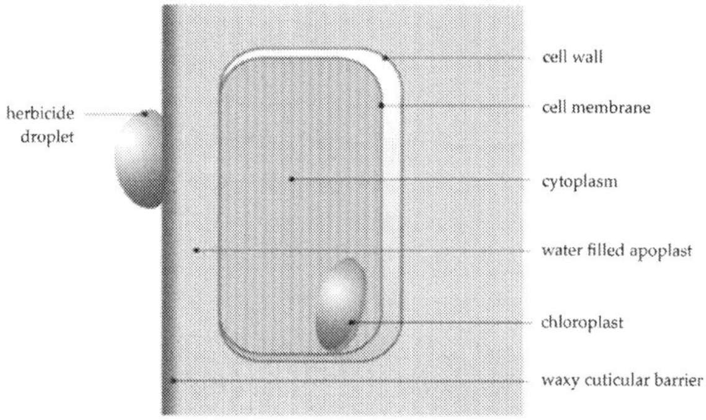

Figure 2.7: Herbicide effectivity should move from the outside leaf that passes through different passage to enter into the target site of the cell.

BIOLOGICAL PROPERTIES OF FUNGICIDES

Fungicides are type of pesticide applied to crops attacked by fungi and molds especially in the roots and leaves, even the stored foods and seeds could be damage by this fungus that attached itself even in the skin of human beings and animals. Fungi are pest that can damage crops because of the presence of a toxin known as " mycotoxins". Example is a fungus, *Aspergillus flavus* that can affect nuts such as peanuts and grains in corn. This fungus produces aflatoxin, a compound that can cause liver disease and in some situations, liver cancer. Another naturally occurring grain fungus produces an ergot alkaloid that can cause hallucinations and fungus that attach itself in the leaves of tobacco, *Peronosporatabacina as shown in figure 2.6*

Figure 2.8: A blue mold (Peronosporatabacina) attack the leaves of tobacco (R.J. Reynolds Tobacco Company Slide Set, R.J. Reynolds Tobacco Company, Bugwood.org)

In the early part of time, fungicides were mercury based compounds, sulphur and copper sulphate substances. During 1940's to 1950's hexachlorobenzene becomes one of the main components in a synthetic fungicide which was widely used to protect the roots due to fungal presence in roots and leaves. Mercurial compounds were also applied to seed grains to protect them from fungus in the soil.Wherein these chemicals may caused severe illness with people who can ingest or eat grains treated during the planting periods of this crops. These two fungicides are now hardly used due to its toxic effect and been replaced by less potent ones, but careful harvest and storage procedures for seeds are necessary to prevent potential contamination of food supplies.

The prime need for fungicides in crop production, seed and crop storage can be reduced by controlling environmental conditions that encourage the growth of fungi. The Integrated Pest Management in every country should implemented strictly inorder to reduce the need to use dangerous fungicides on growing plants.

Another way to eliminate or control the use of fungicides is to substitute this with biofungicide. Biofungicides are microbial pesticides that uses a naturally occuring substances by the use of biochemical pesticide method that are approved for organic production. Biofungicides are widely used by organic vegetable growers to control selected foliar and soilborne diseases of vegetable crops (see Table 2.1).

Application of biofungicides is a kind of treatment that controls the target disease in plants as long as it is applied beforehand the plant disease had contaminated the plant (Francis and Keinath, 2010). In the case of a biofungicide, the Latin name of the microbe that it contains is the generic name of the fungicide.

Table 2.1: List of biofungicides (biological) used to control selected vegetable crop diseases (Francis and Keinath, 2010)

Product	Active Ingredient	Disease	Treatment Site
Ballad	*Bacillus pumulus*	Several (Foliar)	Foliar
Bio-Save	*Pseudomonas syringe*	Post-harvest	Irish and sweet potatoes in storage
Contans	*Coniothyrium minitans*	White Mold	Soil applied
Kodiak	*Bacillus subtilis*	*Phytium,Rhizoctonia, Fusarium*	Seed treatment, beans only
Mycostop	*Streptomyces griseoviridis*	Several	Greenhouse, Soil applied
Regalia	Plant extract	Powdery mildew	Foliar
Root shield granules,Root shield WP	*Trichoderma harzianum*	*Pythium, Rhizoctornia, Fusarium*	Soil applied
Serenade	Bacillus subtilis	Powdery mildew,other foliar diseases	Foliar
T22-HC	*Trichoderma harzianum*	*Pythium, Rhizoctornia, Fusarium*	Soil applied
Surround	Kaolin	Powdery mildew	Foliar
Trilogy	Neem Oil	Powdery mildew	Foliar
Actinovate AG	*Streptomyces lydicus*	*Fusarium, Rhizoctonia, Pythium, Phytophthora*	Soil applied Foliar
SoilGard	*Gliocladium virens*	Damping off	Greenhouse-transplants, soil applied

The USEPA identified the different application of fungicides that can either be subchronic, chronic, developmental, reproductive toxicity, genotoxicity and oncogenicity studies for federal registration of Eco GuardTM Biofungicide. Instead, the USEPA used reports from the scientific literature to evaluate this product.

Fungicides can either be chemicals or biological in nature that can prohibit the growth of fungi or fungal spores. Recent application of fungicides do not kill fungi, they simply inhibit growth in a period of time probably for weeks. Fungi applied to agriculture has negative effect that may result in critical losses of yield, quality and profit. Fungicides are used both in agriculture in controlling fungal infections in animals. Chemicals used to control a psuedo fungi known as oomycetes that are also referred to as fungicides as oomycetes use the same mechanisms as fungi to infect plants (Latijnhouwers et al., 2000).

Fungicides can be transmitted either by contact or transfer in contact fungicides that are not taken up into the plant tissue, & only protect the plant where the spray is deposited.The translaminar fungicides redistribute the fungicide from the upper and then sprayed the leaf surface to the lower part. The unsprayed surface will be applied with systemic fungicides that are taken up and redistributed through the xylem vessels to the upper parts of the plant. In this process a new leaf growth is protected for a short period.

Most fungicides are being retailed and sold in the market in liquid form. The most common active ingredient is sulfur, present at 0.08% in weaker concentrates, and as high as 0.5% for more lethal fungicides. Fungicides in a powdered form contains usually about 90% sulfur and are very potent. Other active ingredients in fungicides include organic oils from neem, rosemary and jojoba with the addition of the bacterium *Bacillus subtilis*. Fungicide residues have been found on food for human consumption, mostly from postharvest treatments (Brooks & Roberts, 1999).

Finally some fungicides are destructive to human health, such as vinclozolin, which has now been removed from use (Hrelia, 1996), FCX and DFB that are used as pesticides to control pests and they have many side effects on natural non-target organisms (Rouabhi et al., 2009).

Biological Properties of Rodenticides

Rodenticides are type of pesticide eliminate by killing small mammals especially rats and mice. Some rodenticides works as anticoagulants that inhibits blood-clotting which are often used to eradicate rat populations that are considered as pests by farmers. One of the first anticoagulant rodenticides applied that can minimize or totally controls rat population was warfarin, which is related to plant-derived coumadin(from spoiled sweet clover).

In the 1950's resistance to rodenticide warfarin had occured which prompted scientists to develop more stronger and effective anticoagulants, which are termed second-generation anticoagulants. Other kinds of rodenticides are fluoroacetic acid and zinc phosphide which are both toxic and thiourea-based compounds.

Some problems that can lead to the application with rodenticides can also harm the wildlife by mistake, when baits for rodents and pellets would accidentally eaten by wildlife. Thus, wildlife such as birds, wolves and forms of wildlife would be harmed by accidentaly eating the food set as food for rodents. This might poisoned the predators and other wildlife due to mistaken food. The primary alternative to using chemical rodenticides is trapping.

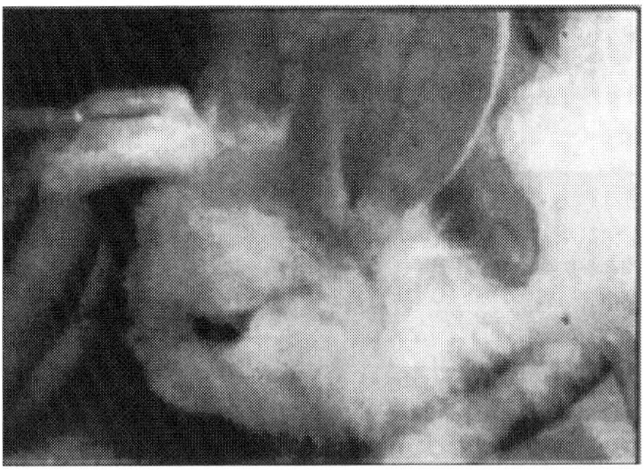

Figure 2.9: An experimental mice feed with pesticide.

Many different active chemical compounds comprising a registered as rodenticides than can be grouped together according to how they work. Many rodenticides stop normal blood clotting or known as anticoagulants. Example of this rodenticides are the Bromadiolone, chlorophacinone, difethialone, brodifacoum, and warfarin are all anticoagulants. There are a number of rodenticides that are not anticoagulants, and these work in different ways.Table 2.2 shows the classification of different rodenticides with the presence of the chemical compounds in terms of its toxicity.

Table 2.2: Classification of Rodenticides in terms of its Toxicity

Active Chemical compounds	Oral	Inhalation	Dermal	Primary Eye Irritation	Primary Skin Irritation
Warfarin[3,6]	Moderate - toxicity	Not significant	Not significant	No data	No data
Chlorphacinone[4]	High toxicity	High toxicity	High toxicity	Non-irritating	Non-irritating
Diphacinone[4]	High toxicity	High toxicity	High toxicity	Moderate irritation	Slight irritation
Bromadiolone[4]	High toxicity	High toxicity	High toxicity	Low irritation	Minimally irritating
Difethialone[3]	High toxicity	High toxicity	High toxicity	Mild irritant	Non-irritating
Brodifacoum[4]	High toxicity	High toxicity	High toxicity	Minor irritation	Mild irritant
Bromethalin[4]	High toxicity	High toxicity	Moderate toxicity	Slight irritation	Non-irritating
Cholecalciferol[3]	High toxicity	Very low toxicity	Low toxicity	No data	No data
Zinc phosphide[5]	High toxicity	High toxicity	Low toxicity	Slight irritation	Non-irritating
Strychnine[7]	High toxicity	High toxicity	Low toxicity	Highly irritating	Non-irritating

(Source: Classification categories were modeled after the U.S. Environmental Protection Agency, Office of Pesticide Programs, Label Review Manual, Chapter 7: Precautionary Labeling.https://www.epa.gov/oppfead1/ labeling/lrm/chap-7)

BIOLOGICAL PROPERTIES OF MOLLUSCICIDES

Slugs or snails are coming from the phylum Mollusca – the same phylum shared by oysters and clams – and the class, Gastropoda. There are three families into which slugs are divided: Arionidae, Limacidae, Milacidae (Glen et al.,1996). The most common slugs in the Midwest is *Deroceras reticulatum*, the grey garden slug (Shetlar, 2000).

Slugs, snails or molluscicides are often described as a "snail without a shell". Like snails, they have a soft, unsegmented body 2.5 – 20 cm long, depending on species. It is covered with a layer of sticky mucus that helps them to move about. Because of this sticky mucous layer, slugs require conditions and thus seek out dark, cool areas, emerging from their hiding places evening or at night.

A snail's body consisted of different parts such as the head, neck, mantle, tail, and foot. It has two pair of tentacles - one large pair with eyes on extensible eye stalks and a small that is used for smelling. Its mouth is located in the center of head wherein it's lower tentacles and is highely equipped with chewing mouthparts and a horn-like "radula" used to rasp away at food. Most of its internal parts are hiddent in it's mantle. The slug moves by way of muscular "foot" and leaves shiny slime trail behind. Colors range from almost black to orange (Nielsen, 1997).

Slugs are hermaphrodites that could either cross or self-fertilize and has a mating seasons around August to October by laying egss in about thirty to forth days period. Slugs overwinter as adults or nearly mature can generally survived in year at outdoors and longer in a greenhouse and can be able to lay 20-100 eggs several times per year. Eggs are laid near soil surface which is a little bit misty and in a moist area. They are held together with a sticky substance that crawls and feed immediately upon hatching and eggs can remain viable over extended time in a dry place and will hatch in a damp area (Nielsen, 1997; Sheltar, 2000). During the spring an adult slug can survive and would have 90,000 grandchildren and 27 million great grandchildren (Symondson, 1996).

Molluscicides a type of pesticide that used to control slugs and snails reproduction. Molluscs are a group of invertrebate animals thatinclude shellfish, cephalopods such as squid and octopus and snails.The most commonly used active ingredient found in molluscicides are metaldehyde compounds which disrupts the gastric organs of slugs and snails, causing death.

Molluscicides products is often manufactured as brightly colored pellets, which should be taken care of because if left unattended this will be dangerous to children because of its attractive color to attract snails in eating it but deliterous to children, wildlife, and pets. Some manufacturers have added a bitter agent to make the products unpalatable. Alternatives to synthetic molluscicides using traps as barriers applied to gardens inorder to make it less attractive to slugs. Slug and snail baits containing iron phosphate as the active ingredient are also available and are less toxic than metaldehyde.

REFERENCES

1. Blondell, J. (1997). Epidemiology of pesticide poisoning in the United States, with special reference to occupational cases. Occup. Med., State of the Art Rev., Vol., 12:209–220.
2. Brooks, G.T. & Roberts, T.R. (1999). Pesticide Chemistry and Bioscience. Published by the Royal Society of Chemistry.
3. Burpee, L. (2006). Integrated disease management, an introduction to Fungicides. Courses support.
4. Costa, L.G. (1997). Basic toxicology of pesticides. Occup. Med. State of the Art Rev. Vol., 12:251–268.
5. Edwards, I.R., Ferry, D.G. & Temple, W.A. (1991). Fungicides & related compounds, In: Handbook of Pesticide Toxicology. Hayes, W.J. & Laws, E.R., Eds. Academic Press, New York, NY,. Vol., 3, pp: 1409–1470
6. Envirofacts, (2005). CYPROCONAZOLE The Active Ingredient in Alto® One of the Active Ingredients in Quadris Xtra®. Syngenta Crop Protection Inc., Greensboro, NC 27419-8300.
7. Extoxnet, (1994). Pesticide Management Education Program, Cornell University, NY. FAO/WHO. (1968). Evaluations of some pesticide residues in food. FAO/PL: 1967/M/11/1; WHO/Food Add./68.30.
8. Fetzer, Jody, "Slugs in the Hosta", Minnesota Landscape Arboretum News, March/April 1994, volume 13, no. 2.
9. Fetzer, Jody, "Slugs!", University of Minnesota Landscape

Arboretum information flyer. Provided by Fetzer.

10. Fishel, F.M. (2005). Pesticide Toxicity Profile: Triazole Pesticides. University of Florida, IFAS extension. PI68.
11. Foster, and Smith, (2010). Germs: Viruses, Bacteria, and Fungi. Veterinary & Aquatic Services Department Foster & Smith, Inc. Wisconsin. USA.
12. Francis, R. and Keinath, A. (2010). Biofungicides and chemicals for managing diseases in organic vegetable production. CLEMSON Cooperative extension. Information leaflet 88.
13. Glen, D.M., Wilson, M.J., Hughes, L., Careeg, P., and Hajjar, A. "Exploring and Exploiting the Potential of the Rhabditid Nematode *Phasmarhabditis hermaphrodita* as a Biocontrol Agent for Slugs", pp. 271-280, Slug and Snail Pests in Agriculture: Proceedings of a symposium held at the University of Kent, Canterbury, September 24-26, 1996. British Crop Protection Council, Surrey, UK.1996.
14. Hahn, Jeff, "Slugs in the Garden", University of Minnesota Extension Service, bulletin 446. 2000. Cited April 28, 2001. Available online http://www.extension.umn.edu/info-u/plants/BG446.html
15. Hahn, Jeff, and Fetzer, Jody, "Slugs in Home Gardens", 2000, University of Minnesota Extension Service bulletin FS-7561-GO, Regents of the University of Minnesota. Cited April 28, 2001. Available online http://www.extension.umn.edu/distribution/horticulture/DG7561.html
16. Harding, W.C. (1979-80). Pesticide profiles, part two: fungicides and nematicides. Univ. Maryland, Coop. Ext. Service Bull. 283, 22 pp.
17. Hayes, W.J and Vaughn, W.K. (1977). Mortality from pesticides in the United States from 1973 to 1976. Toxicol. Appl. Pharmacol. Vol., 42: 235–252.
18. Howard, P. H., (1989). Handbook of Environmental Fate and Exposure Data for Organic Chemicals: Pesticides. Lewis Publishers, Chelsea, MI., 4-20
19. Hrelia, R. (1996). The genetic and non-genetic toxicity of the fungicide. Vinclozolin Mutagenesis. Vol., 11: 445-453.
20. Latijnhouwers, M.; de Wit, P.J. and Govers, F. (2000). Oomycetes

and fungi: similar weaponry to attack plants. Trends in Microbiology. Vol., 11: 462-469.

21. Litovitz, T.A.; Felberg, L. & Soloway, R.A. (1994). Annual reports of the American Association of Poison Control Centers. Toxic exposure surveillance system. Am. J. Emerg. Med. Vol., 13: 551–597.
22. Marrs, C.T. & Ballantyne, B. (2004). Pesticides toxicology and international regulation. John Wiley and Sons, Ltd. 554 pages.
23. Nielsen, G.R., "Slugs and Snails", University of Vermont Extension, bulletin EL-14, January 1997. Cited April 8, 2001. Available online http://ctr.uvm.edu/ctr/el/el14.html
24. Novozymes Biologicals, Inc. (2007). Roots EcoGuard biofungicide. www.novozymes.com/roots.
25. Office of Pesticide Programs, (1996) List of Chemicals Evaluated for Carcinogenic Potential, US EPA, Washington, US.
26. OHS Database. 1994. Occupational Health Services, Inc. 1994. MSDS for Vernolate. OHS Inc., Secaucus, NJ.
27. Parent, R.A.; Caravello, H.E. and Hiberman, A.H. (1992b). Reproduction study of Acrolein on two generations of rats. Fund. Appl. Toxicol., Vol., 29: 228–237.
28. Parent, R.A.; Caravello, H.E. and Long, J.E. (1992a). Two-year toxicology and carcinogenicity study of acrolein in rats. J. Appl. Toxicol., Vol., 12: 131–139.
29. Phillips, S.D. (2001). Fungicides and biocides. In: Clinical Environmental Health and Toxic Exposures, Sullivan, J.B. & Krieger, G.R., Eds. Lippincott Williams and Wilkins, Philadelphia, 2nd Eds. pp:1109–1125.
30. Rouabhi, R.; Djebar, H. & Djebar, M.R. (2009). Toxic Effects of Combined Molecule from Novaluron and Diflubenzuron on Paramecium caudatum. Am-Euras. J. Toxicol. Sci. Vol., 1(2). September 2009. 74-80. ISSN: 2079-2050; EISSN: 2079-2069.
31. Royal Society of Chemistry (1983). The Agrochemicals Handbook. The University, Nottingham, England.
32. Serafini, P.M. (2003). Bacillus licheniformis SB3086 NYS DEC Letter - New Active Ingredient Registration 8/03. New York State

Department of Environmental Conservation.

33. Shetlar, David J. and Herms, Dan, *Insect and Mite Control on Woody Ornamentals and Herbaceous Plants*, 1997-1998, pp. 72-73, The Ohio State University, Columbus, Ohio

34. Shetlar, David J., "Slugs and Their Management", 2000, Ohio State University Extension, bulletin HYG-2010-95. Cited April 8, 2001. Available online http://www.ag.ohio-state.edu/~ohioline/hyg-fact/2000/2010.html

35. Speiser, B., Glen, D., Bohan, A., Aalten, M., Hughes, L., Rowcliffe, H., Gwynn, R., Gupta, A., Ester, A., van Rozen, K., Huiting, H.F., Davies, K., Denholm, C., Castillejo, J., Iglesias, J., and Coupland, J "Novel Slug Control: Combined efforts to devise strategies of slug control for organic and integrated farming". Poster presented at the 13th International IFOAM Scientific Conference, Basal Switzerland, August 2000. Cited April 28, 2001. Available online http://www.slugcontrol.iacr.ac.uk/speiseretal.html

36. Symondson, Dr. Bill, "Slug Control", University of Wales, Cardiff, United Kingdom. October 1996. Cited April 27, 2001. Available online http://www.oxalis.co.uk/_

37. Thomson, W.T. (1990) Agricultural Chemicals, Book IV: Fungicides. Thomson Publications, Fresno, CA.

38. Tomlin, C., (1994). The Pesticide Manual, 10th Edition, British Crop Protection Council/Royal Society of Medicine.

39. U.S. Environmental Protection Agency, (1993). Office of Pesticides. TOX Oneliners Triforine. April, 1993.

40. U.S. National Library of Medicine, (1995). Hazardous Substances Data Bank. Bethesda, MD., 4-5

41. Whitehead, R. (1996). The UK Pesticide Guide, British Crop Protection Council/CAB International.

42. Willoughby, OH. NAS, (1987). Regulating Pesticides in Food. The Delaney Paradox. Report of Committee on Scientific and Regulatory Issues. Unlikely Pesticide Use Patterns. National Academy of Sciences, National Academy Press, Washington, DC

43. World Health Organization, (1993). Benomyl, Environmental Health Criteria No 148, Geneva, Switzerland.

44. World Health Organization, (1994). Data sheet on benomyl, WHO/PCS/94.87, Geneva. Worthing, C.R. (1979). The Pesticide Manual: A World Compendium, 6th Eds. The British Crop Protection Council, Croydon, England. 655 pp.
45. Worthing, C.R. (1983). The Pesticide Manual: A World Compendium. Seventh edition. Published by The British Crop Protection Council.

3
CHAPTER

EFFECTS OF PESTICIDE ON HUMAN HEALTH

INTRODUCTION

The purpose of pesticides is to eradicate certain pest because their mode of action is not specific to one species, they often kill or harm organisms other than pests, including humans. The World Health Organizationhas counted about 3 million cases of pesticide poisoning each year and up to 220,000 deaths, primarily in developing countries.

Pesticides application is often not very specific and accurate, and sometimes unintended exposures occur to other organisms in the general area where pesticides are applied. Children and even any young and developing organisms are often vulnerable to the harmful effects of pesticides. Even very low levels of exposure during development may have adverse health effects.

Exposure to pesticide can cause a wide range of neurological health effects such as memory loss, loss of coordination, reduced speed of response to stimuli, reduced visual ability, altered or uncontrollable mood and general behavior, and reduced motor skills. Some symptoms are not so adverse and may not be recognized by the medical community as a clinical effect. Possible health effects may include asthma, allergies, and hypersensitivity, and pesticide exposure is also linked with cancer, hormone disruption, and problems with reproduction and fetal development.

Pesticide can neither contain both "active" and "inert or inactive" ingredients. Active ingredients are what kill the pest, and inert ingredients help the active ingredients to work more effectively. These "inactive" ingredients may not be tested throughly as active ingredients and are seldom disclosed on product labels. Solvents, which are inert ingredients in many pesticide formulations which could be dangerous or toxic either be poisonous through inhalation or by contact through skin. As seen in figure 3.1 showed the global consumption of pesticide in the whole continent from year 2000 to 2012 shows that there is an increase of pesticide used especially in Asia and the highest consumption percentage coming from the European consumers and farmers.

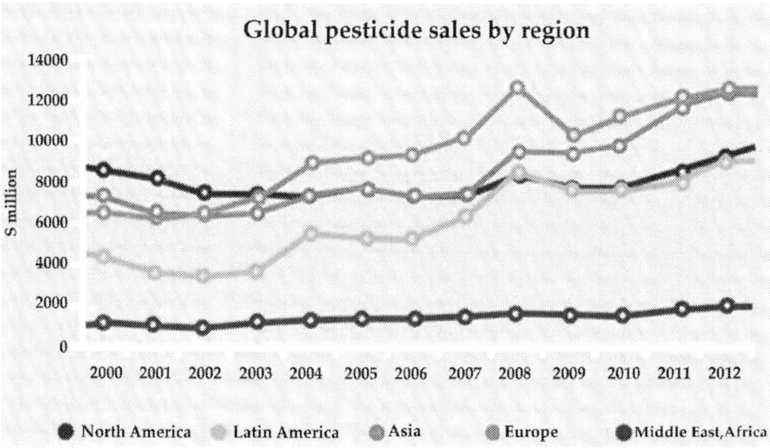

Figure 3.1: The global pesticide consumption in different continent starting year 2000 to 2012.

Children are at greater risk from exposure to pesticides because of their weaker resistance especially due to relative to their size, children eat, drink, and breathe more than adults. Children bodies and organs are still in the development stage such as growing rapidly, which also makes them more susceptible; in fact, children may be exposed to pesticides even while in the womb.

The application of pesticides are intended to harm only the target pest, but if not used correctly, they can also harm people or the environment. Pesticide in the environment is not necessarily a problem, but due the presence of the chemical residue exposure of the non-target may also harm other organisms. As with all toxic substances, whether the exposure causes harm will depend on the right dosage, how someone is exposed, how sensitive an individual may be to that toxin, and the toxicity of the pesticide involved.

People can be exposed to pesticides in three ways:

- Breathing (inhalation exposure).
- Getting it into the mouth or digestive tract (oral exposure).
- Contact with the skin or eyes (dermal exposure).

Pesticides has three different routes as an entry to the body,they could enter through inhalation exposure by breathing it directly that contains the gas or moist content in air containing pesticide as a in gaseous form such vapor, as an aerosol, or on small particles like dust. Oral exposure happens when one can accidentally ingest it through eating as food or drink water containing pesticides. Dermal exposure can happen when skin is exposed to pesticides. Thus, it can cause irritation or burns and in most serious cases, your skin can absorb the pesticidedirectly to human body, causing other health effects. Some pesticides evaporate more easily than others so they are more likely to be inhaled. Some break down quickly on surfaces; others last longer.

Application of pesticide in a liquid spray can be spread easily than a dry granules. A pesticide sealed in a trap is less likely to be eaten accidentally by a child or a pet. A dry pesticide can leached into the soil can be dangerous to ground-water, but is not as likely to drift through the air. All these factors affect the potential risk of human exposure and are considered when DPR makes rules for pesticide use.

EFFECT OF INSECTICIDES IN HUMAN HEALTH

Organochlorines

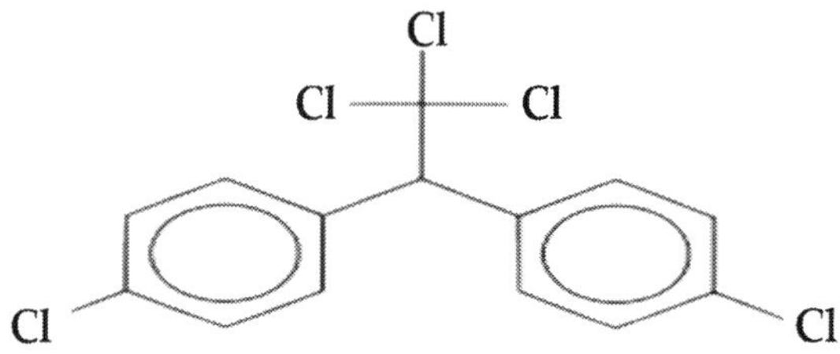

Figure 3.2: The chemical structure of organochlorines.

Organochlorine the main component of insecticides results a universal problem due to the presence of contaminants in the environment and has side effects on human health. This constitutes a public health concern due to the compound organochlorines have been linked with cancer, asthma, diabetes, and growth disorders in children (Rosman,Y etal.,2009).

The effect organochlorines to a person occur via ingestion of contaminated food or water, inhalation of vapor, and absorption through the skin. Occupational and other domiciliary exposures are also possible and even food intake with small traces of organochlorines results in bioaccumulation of these chemicals in the human body (Snedeker S., 2006).

Organochlorines have the structural conformation that hey all contain a cyclodiene ring. The different body parts such as lungs, gastrointestinal tract, and skin can absorb all these compounds. Although some animals may absorbs approximately 10% of the applied dose, lipid solvents increase dermal penetration(Ellenhorn & Barceloux, 1988) because of this it has a high risk of intoxication in the case of workers who apply these products in crops without proper protective equipment.

Organochlorine compounds have a unique chemical structure and especially in their physical properties such as polarity and solubility. These fat-soluble compounds persist both in the human bodies and the environment. Consequently, researchers and regulatory agencies have banned several organochlorines (Barr & Needham,2002).

Some symptoms of organochlorines when ingested are dizziness, headache, anorexia, nausea, vomiting, malaise, dermatitis, diarrhea, apprehension, excitement, irritability, gait disorders, too much sweating, altered reflexes, muscle weakness, uncontrollably tremblings, spasms, mental confusion, anxiety, seizures, coma, and death. The carcinogenicity of this class of compound affixed to polychlorocyclodiene compounds that form epoxides during their biotransformation. Because organochlorines have a long half-life, these levels in the serum constitute a marker of exposure to these pesticides (Ellenhorn & Barceloux, 1988).

Treatment for organochlorines accidental ingestion, it is necessary to maintain the vital functions, administer diazepam and phenobarbital by slow injection, to control seizures, and be sure to monitor the air passage closely. Lorazepam can be a substitute as an alternative to diazepam. Ion exchange resins can also be administered orally. Arrhythmias that damage the myocardium rarely occur. Lidocaine is the treatment of choice (Simpson & Schuman,2006).

Higher level of ingestion of organochlorineinsecticides can cause a loss of sensation around the mouth, hypersensitivity to light, sound, and touch, dizziness, tremors, nausea, vomiting, nervousness, and confusion.

During the year of 1975, about 70 workers manufacturing Kepone, an organochlorine insecticide, in Hopewell, Virginia, developed a variety of neurological symptoms, the most prominent incident ever known as the "Kepone shakes." The workers' symptoms started about 30 days after their first exposure to Kepone. Subsequent testing also revealed decreases in sperm count and motility. In 1976, Kepone was discontinued and substituted with organophosphates.

Organophosphates

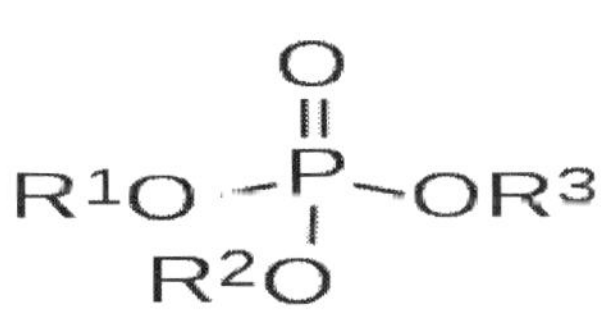

Figure 3.3: The chemical structure of organophosphates.

Organic compounds containing phosphorus, is associated with the so- called organophosphorus compounds (OP), that has been found in the application pesticides and war against the use of this gases and its synthesis during the year of 1937. Organophosphorous contains carbon derivative from phosphorous acid. Their primary structure may vary depending on whether they bear sulfur or oxygen double bonds. X is a group of the general structure that separates when the compound binds to acetylcholinesterase (AChE) (Barr & Needham,2002).

The variations on the general structure of organophosphorus is possible and can be subdivided these compounds into phosphates, phosphorothioates, phosphoramidates, and phosphonate, for example. The difference in the structure between these compounds results in peculiar characteristics regarding OP metabolism and toxicity. Some representatives of this class of pesticides are Diazinon, Malathion, and Paration (Ellenhorn & Barceloux, 1988).

Outer skin, conjunctiva, gastrointestinal tract, and lungs rapidly absorb most OP compounds. Cytochrome P_{450} isozymes metabolize these chemicals in the liver, which sometimes generates metabolites that are more toxic than the parent compounds (Costa,L., 2008). One example is the oxon form, which may bind to cholinesterase or undergo hydrolysis to a dialkyl phosphate and a hydrolyzed organic moiety specific to the pesticide (Barr & Needham,2002) . Most OP are polar and hence water soluble. Their metabolites arise 12 to 48 h after exposure. However, a few compounds, such as dichlofenthion, possess high partition coefficients, which culminates in long-lasting symptoms (Davies, etal., 1975).

Too much exposure to carbamate and organophosphate shows an early symptoms of too much intake of acetylcholine, increase in perspiration and salivation, diliated pupils, stomache pain, vomiting, diarrhea, increase and decrease in blood pressure, lost of muscle coordination. Symptoms slowly disappears if the intake of acetylcholine levels ends and back to normal activity.

Some effects of organophosphates may be associated with delayed neurological reaction characterized by muscle weakness in the legs and arms. During Prohibition, people consumed a homemade alcoholic drink from a Jamaican ginger which was contaminated with the organophosphate triorthocresyl phosphate (TOCP). About 20,000 people were affected by a condition called "Ginger Jake paralysis" that was found in some researches that can has it's effect in the reproduction of animals and the

US government required that organophosphates be tested for delayed effects during the registration process. The lethality of human's in terms of organophosphates caused a decline in their use and spurred the search for new alternatives.

Symptoms of pesticides intoxication can be associated with some syndromes like the muscarinic syndrome, this is due to the action of acetylcholine on the smooth muscle, heart, and exocrine glands that enlarges bronchial secretion, tearing, and sweating; stops the movement of the gastrointestinal tone to cause nausea, vomiting, and diarrhea; and elicits urinary incontinence, bronchospasm, miosis, and bradycardia.

For instance in nicotine syndrome, in which acetylcholine accumulates at the motor nerve endings in the autonomic ganglia and causes tremblings, spasms, hypertonicity, hyperreflexia, paralysis, or muscle weakness and enhances the sympathetic autonomic ganglia, to promote tachycardia, pallor, hyperglycemia, and hypertension.

In addition this could also affect the central nervous system (CNS) that includes some actions such as anxiety, headache, dizziness, ataxia, sleep and memory disorders seizures, tremors, respiratory depression, and coma. Some organophosphorus containing compounds can still have teratogenic potential and mutagenic effects. Laboratory diagnosis of this syndrome involves determination of cholinesterase activity (Paudyal ,2008).

Treatment on the organophosphorus poisoning should be maintained inorder to lessen its effect in the damage of the vital functions and assess cholinesterase levels in the red cells and lower the level of the pseudocholinesterase present in the plasma, before therapy. It should be noted to avoid the use of parasympathomimetic agents that can enhance the activity of the anticholinesterase action. Treatment should start with atropine, which acts as a competitive muscarinic anticholinergic agent, together with pralidoxime, until complete control of the symptoms. After the medication atropinization, administration of furosemide prevents pulmonary congestion, whereas administration of benzodiazepines controls seizures (Thiermannetal., 2013).

Carbamates

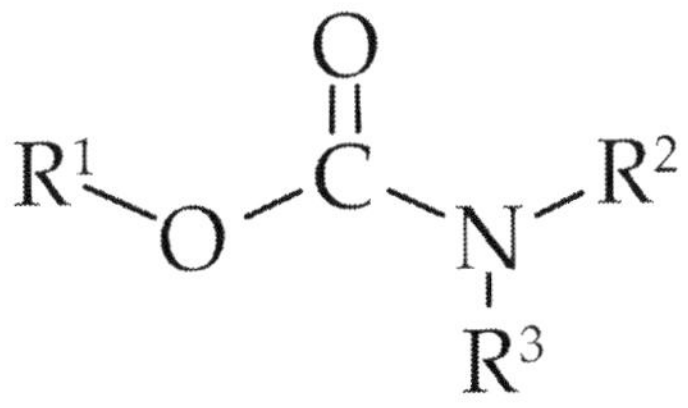

Figure 3.4: The chemical structure of carbamates.

Carbamates, a type of insecticide that produces healthwise signs and symptoms of cholinergic excess that resembles the signs associated with organophosphate toxicity, except that the effects are more reversible and less severe(Barr & Needham ,2002). The mechanisms underlying carbamates poisoning involve carbamylation of the active site of acetylcholinesterase, which inactivate this essential enzyme in the nervous system of humans and other animal species (Handbook of Pesticide Toxicology,2001).

Carbamates reactions with acetylcholinesterase has the action mechanisms with organophosphorus with the same enzyme. However,the reaction activiation of the carbamylated enzyme by hydrolysis has faster metabolic action compared with the activation energy when the presence of phosphorylated enzyme, with reversal of inhibition typically occurring half an hour or less after exposure (Toxicology Letters 2009) and nevertheless, reports on cases of neuropathy after poisoning exist (Marrs & Maynard, 2013).

Organisms can easily metabolized carbamates compounds through the following parts of the body such as lungs, gastrointestinal tract, and skin. It is just good enough that carbamates could poorly penetrate the blood-brain barrier.This is good action wherein the brain cholinesterases activity minimally and promote fewer CNS symptoms as compared with organophosphates. In effect the reaction in vivo hydrolysis of the carbamate-cholinesterase complex contributes to less severe and less enduring symptoms especially in the brain neurons.

Manifestations of carbamates poisoning are the following ; the miosis, salivation, sweating, tearing, rhinorrhea, behavioral change, abdominal pain, vomiting, diarrhea, urinary incontinence, bronchal spasms , dyspnea, hypoxemia, bradycardia, bronchial secretions, pulmonary edema, respiratory failure, drop in body temperature, incoordination, lip tingling, tremors, and seizures. Another manifestations may inculde muscle pains, twitching, muscle weakness (including respiratory muscles), paralysis, tachycardia, and hypertension (Rosmanetal., 2009).

Carbamates poisoning can be treated by maintaining the treatment with the vital functions and to avoid some crucial problems and this could be applied to parasympathomimetic agents, because they may increase the anticholinesterase activity. Medications should begin with the intake of atropine, followed by administration of furosemide, only if necessary. If poisoning is cause by the presence of pure carbamates only, it is not necessary to administrate pralidoxime, except in cases that these carbamates are associated with OPs (Ellenhorn & Barcelou, 1988).

Pyrcthroids

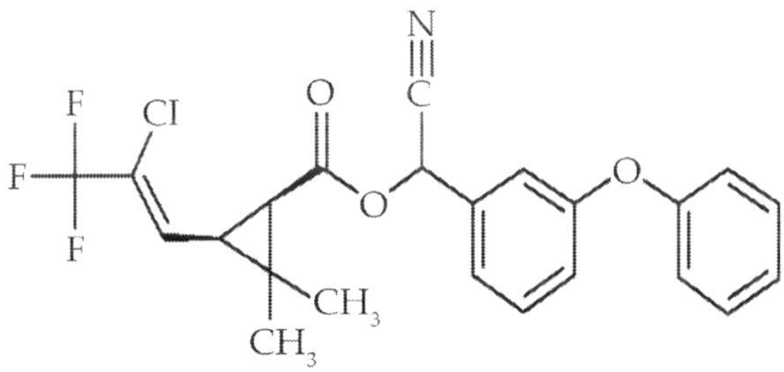

Figure 3.5: The primary chemical structure of pyrethroids and pyrethrins.

Pyrethrins and pyrethroids function mainly as insecticides. Pyrethrins are natural compounds originating from the plant *Chrysanthemum cinerariae folium*. They comprise active agents (pyrethrins I-VI), but pyrethrins I and II has the most active ingredients. These compounds could easily disappear with the presence of light, but synthetic production of pyrethroids around 1950 overcame some disadvantages of natural pyrethrins (Ellenhorn & Barcelou, 1988).

Crude pyrethrum has and effects with the dermal and respiratory allergen, probably due it is to non-insecticidal ingredients. Contact dermatitis and allergic respiratory reactions (rhinitis and asthma) have occurred after exposure to this compound (Reigart &Roberts, 1999).

The most probable usefulness of organophosphates is the synthesis of this into pyrethroids. But, this could cause into different metabolic disorders such as too much excitation, being aggressive, the uncoordinated movement, uncontrollable trembling and seizures. High level of exposure to this compound will further lead to allergic reaction to skin which is cause by the presence of pyrethroids and can cause reproductive disorder and the worst is cancer.

Triazines

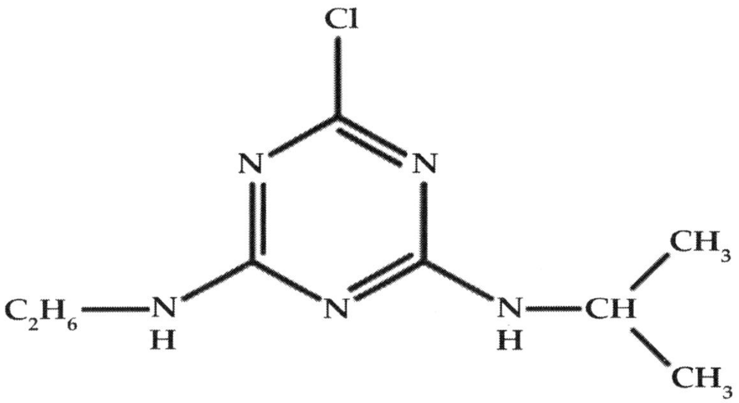

Figure 3.6: The chemical structure of triazines.

One of the largest herbicides commercially available is triazines with a varied uses from agricultural application for fungicides, herbicides and insecticides. Now a days it is even applied in pharmaceutical connection hence it function as alkylating agent as an anticancer chemotherapic cure.

Triazines can be synthesized by reacting cyanuric chloride with the appropriate alkylamine(s) in the presence of acid acceptors. And it shows a good reaction in the degradation process especially in the decomposition of the ring structure. Its herbicidal function is due to the structure of the 2-carbon position: Cl, OCH_3 or SCH_3 are the most common. Of these, Cl results in lower water solubility than either OCH_3 or SCH_3. Substitution at C-4 and C-6 can be more varied without destroying herbicidal activity, but N-alkyl substituents are most common. Thus atrazine, for example, has Cl at the C-2 position, with N-ethyl and N-*iso*propyl substituents at C-4 and C-6.Most of the triazines compounds inhibit the Hill photosynthesis which differs to its photoxicity in different individual triazines which is commonly used in herbicide as a soil sterilant which is needed by the crops.

Geigy chemical company first produce atrazine which is a white crystalline solid with a melting point of 173-175°C and a water solubility of 33 ppm at 27°C an soluble with methanol at 18,000ppm concentration and chloroform at 52,000ppm. It functions as weed control used widely in United States with annual use of about 80 million pounds in the early part of 1980's.

The acute toxicity of atrazine to mammals is low. Its oral LD_{50} to rats is between 2,000 and 3,000 mg/kg, and there is little cumulative toxicity to repeated dosing. Dietary levels of 100 or 500 mg/kg/day caused growth retardation in rats, reportedly because the rats ate less. Symptoms after a single near-lethal dose include respiratory excitation followed, by depression; hypothermia, and motor incoordination. Following a single lethal oral dose, death occurs within 24 hours. Pathologically, lung edema and hemorrhages of numerous organs are seen. The Ld_{50} following dermal dosing in rabbits is greater than 3,000 mg/kg. Atrazine does not appear to cause malformations in rodents, but does cause fetal death at maternally lethal doses in sheep. Fetal deaths have not been reported in any species at doses that were not maternally toxic. These data argue for a negligible level of developmental toxicity.

Phenoxy derivatives

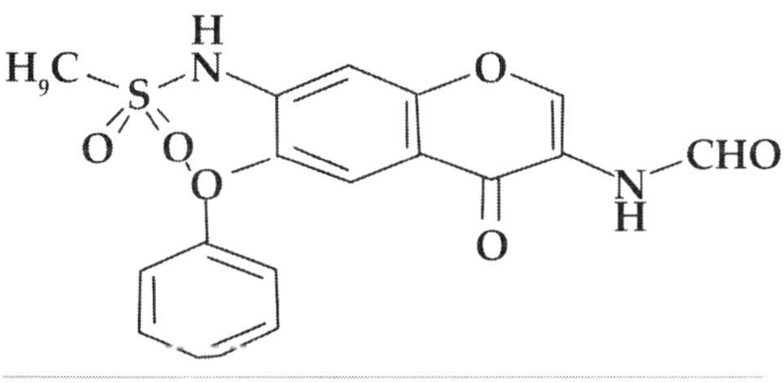

Figure 3.7: The chemical structure of phenoxy derivatives.

The derivatives of phenoxy bearing the structures of an aliphatic carboxylic acid moiety is being attached to a chloride or methyl-substituted aromatic ring. The common type of phenoxy herbicides are the structures of 2,4-dichlorophenoxyacetic acid and 2,4,5-trichlorophenoxyacetic acid.. The combination of these two herbicides in equal proportions affords Agent Orange, a product preliminary applied in the during World war along the jungles of Vietnam, Laos, and Cambodia including in the Vietnam War. Manufacture of phenoxy herbicides often requires co-formulation with ioxynil and/or bromoxynil, which are generally more toxic than the herbicides. Actually a more toxic substances can emerge during the fabrication of some of these herbicides at excessively high temperatures, such as at chlorinated dibenzo dioxin and chlorinated dibenzo furan (Bradberry & Proudfoot, 2004).Because 2,4,5-T contains the highly toxic and persistent 2,3,7,8-tetrachlorodibenzeno-*p*-dioxin along with other chlorinated dioxins and furans, regulatory agencies have banned it for most applications (Barr & Needham,2002) .

Phenoxy salts and esters that can be easily hydrolyze in vivo, so its lethal effect of this derivative will depend mainly on the acid form of the pesticide. Individuals and species may vary differently in terms of phenoxy herbicides elimination. The biological half-life of herbicides in humans reportedly varies from 12 to 72 h (Bradberry & Proudfoot, 2004), but long half-lives occur at large doses and after prolonged exposure (Reigart &Roberts, 1999).

One of the health effect of phenoxy derivatives is on the gastrointestinal tract absorption. The lungs absorb them less, their cutaneous absorption is minimal, and fat does not store them. Phenoxy derivatives has different metabolic actions suchs the mechanisms of toxicity including dose-dependent cell membrane damage, uncoupling of oxidative phosphorylation, and disruption of acetylcoenzyme A metabolism(Bradberry & Proudfoot, 2004).

Dipyridyl derivatives

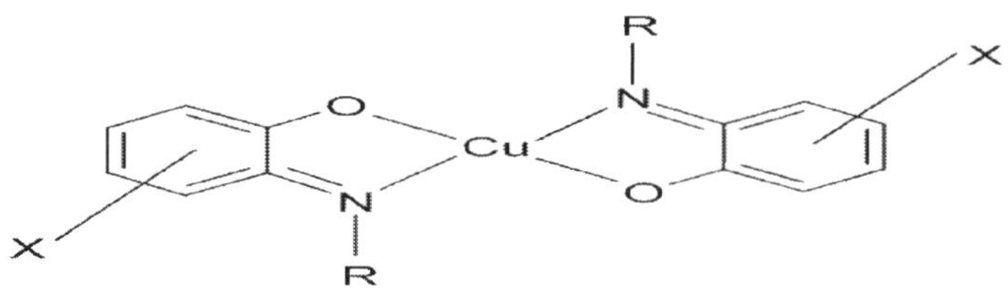

Figure 3.8: The chemical structure of dipyridyl derivatives.

Dipyridine is a hetorcyclic organic compound that contains two rings of pyridine within carbon site in each ring and having a genral formula of $C_{10}H_8N_2$. Every ring constituted with an unsaturated six membered ring into five different carbon atoms in one nitrogen atom.Dipyridine which is use as photosynthesensitizer , luminescent and metal complex materials in chemistry. Farther, dipyridine can be use as a catalyst for polymerization, as luminiescene in spectrophotometric analysis and in polymerization process.

Some dipyridinium molecules, quaternary ammonium compounds, are active ingredients of contact herbicides with non-selective and fast-acting properties. They are poisonous to humans also. They are readily reduced to radical ion which reacts with unsaturated membrane lipids. Pyridinium class plant protection products include cyperquat, diethamquat, difenzoquat, diquat and morfamquat.

Amrinone and milrinone, bipyridine ketone derivatives, are positive inotropic cardiotonic agents with vasodilator effects. They are phosphodiesterase inhibitor that prevents enzymatic breakdown of cyclic AMP by phosphodiesterase. They stimulate calcium ion influx into the cardiac cell. Etoricoxib, a bipyridine non-steroidal antiinflammatory drug, is a COX-2 selective inhibitor.

Glycine derivatives

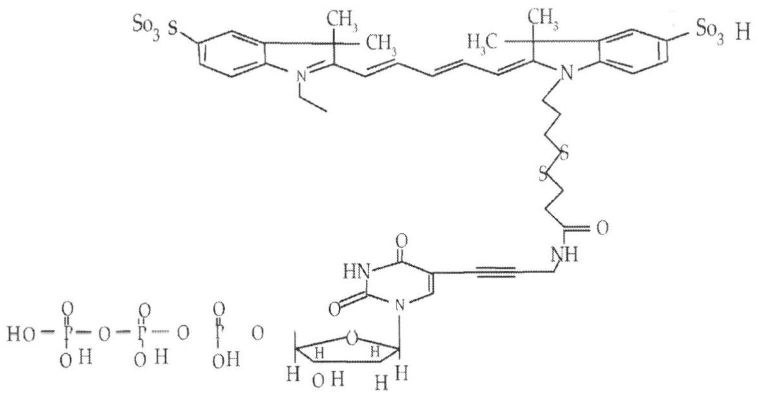

Figure 3.9: The chemical structure of dipyridyl derivatives.

There are two classes of glyphosate (N-phosphonomethyl glycine) and glufosinate (N-phosphonomethyl homoalanine) which is bein sold in the market as the isopropylamine salt (glyphosate) or ammonium salt (glufosinate). This has wide spectrum that controls as a herbicides in controlling a perennial plants. Glyphosate is a compound that has phosphorous in a herbicide used that inhibits the biosynthesis of of tryptophan, phenylalanine, and tyrosine, present in plants, fungi, and bacteria, but not in animals or humans. However, according to literature reports, glyphosate can enter living organisms, including humans, where it exerts various toxic effects.

The mechanisms of toxicity of glyphosate formulations are complicated and the most widely used glyphosate product is Roundup, formulated as a concentrate containing 41% glyphosate which is coming from high concentrations of glyphosate, its metabolites and impurities may reduce acetylcholinesterase (AChE) activity. Although no evidence forsignificant AChE inhibition in mammals in vivo exists (Bradberry & Proudfoot,2004).

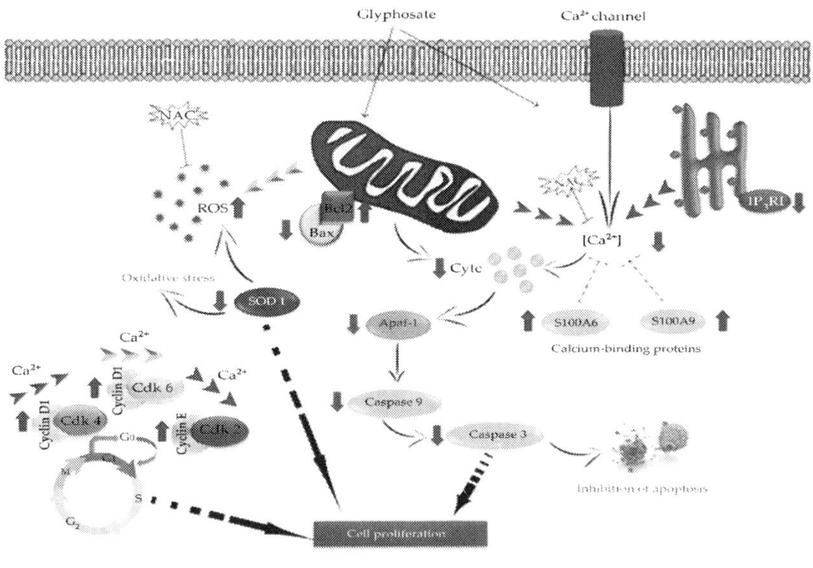

Figure 3.10: The proposed itinerary for glyphosate-induced proliferation in human skin keratinocytes HaCaT cells.

A study published in the *Archives of Toxicology* by Koller and colleagues showed increased in nuclear aberrations after exposure to glyphosate concentrations between 10 and 20 mg/L, which indicated DNA damage (Haliwell & Gutteridge, 2011). In adition, in vitro tests using isolated rat liver mitochondria showed that glyphosate uncoupled the electron transport chain (Hernandez,Parron, Tsatsakis,Requera,Alarcon & Lopez, 2011).

Their are different effect of this compounds ranges from irritation upon local contact (skin, GI) as seen in (figure 3.10) , to hypotension, development of acute renal failure with oliguria, and severe hypoxia and death and the treatment of glycine derivatives poisoning includes maintenance of the vital functions. Hemodialysis is crucial to reduce the amount of toxins normally excreted by the kidney, thereby preventing the impacts on this organ (Hori, Tanaka, Fujisaka & Shemah, 2003).

Dithiocarbamates

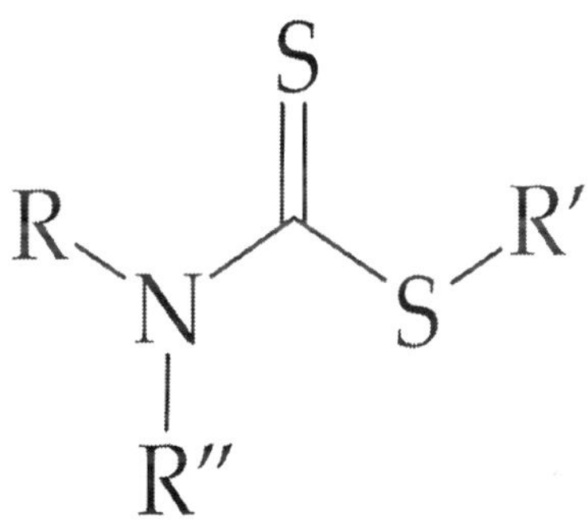

Figure 3.11: The chemical structure of dithiocarbamates derivatives.

Dithiocarbamates comprise two groups: (Abasset al.,2007) dimethyldithiocarbamate and (Abdollahi, Ranjbar,Shadnia, Nikafar & Rezuke, 2004) ethylenebisdithiocarbamate, depending on which metal ion present in the structural formation.Wherein, its name of the different compounds of this class is related to the association of the metal cations; e. g., maneb (manganese), and zineb and ziram (zinc) (Bergasa, 2010; Bradberry,2005).

The slow absorption of these compounds means that they have low acute oral and dermal toxicity. On the other hand, chronic exposure to dithiocarbamates leads to adverse effects due to contact with dithiocarbamate acid or metal ligand (Costa,2008).

The metabolite that arises from dithiocarbamates biotransformation is ethylenethiourea (ETU), that can harm the hormones of thyroid in effect it can lead to thryroid cancer. The enzyme acetalydehyde dehydrogenase which an enzyme that converts acetalydehyde into acetic acid by this enzyme which has the same structure of dithiocarbamates and disulfram.

G

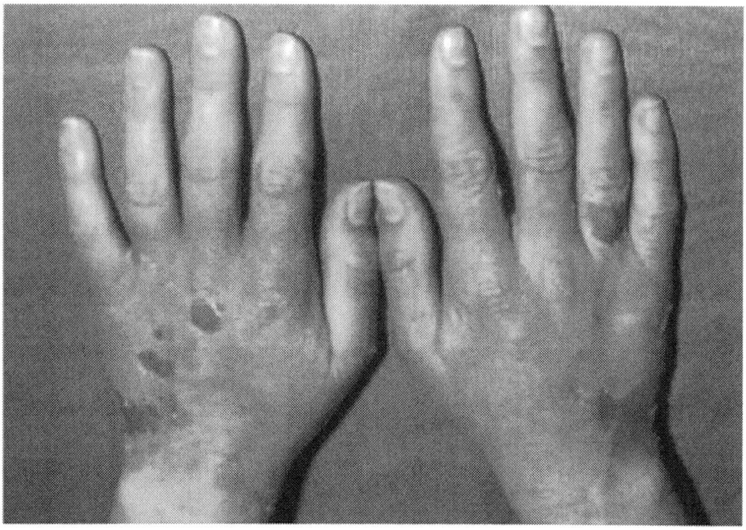

Figure 3.12: Effect of the biotransformation of the toxicity in dithiocarbamates in skin.

Although these products are little toxicity to humans, they are potential precursors of ethylenethiourea, which has carcinogenic and teratogenic action as shown in (figure 3.12).There is no specific treatment for poisoning with this class of compounds, so only maintenance of vital functions and minimization of their absorption (activated charcoal) are necessary.

Others

Others classes of pesticides exist, including the chloroacetanilide commonly used in agriculture. A number of chloroacetanilides, like alachlor, acetochlor, metolachlor, and propachlor are carcinogenic (Inoue,Onodera,Fujita, Fujino, Kikuchu & Endo, 2003). The metabolism of chloroacetanilides most likely proceeds via conjugation with glutathione, as judged from the amount of glutathione-related metabolites in the urine of rats treated with these herbicides. (Jarrard, Delaney & Kennedy, 2003).However, the predicted differences between humans and rats in terms of disposition together with the lower rates of alachlor metabolism in human nasal microsomes have led scientists to question the human relevance of chloroacetanilide olfactory carcinogenicity (Jokanovic, 2009).

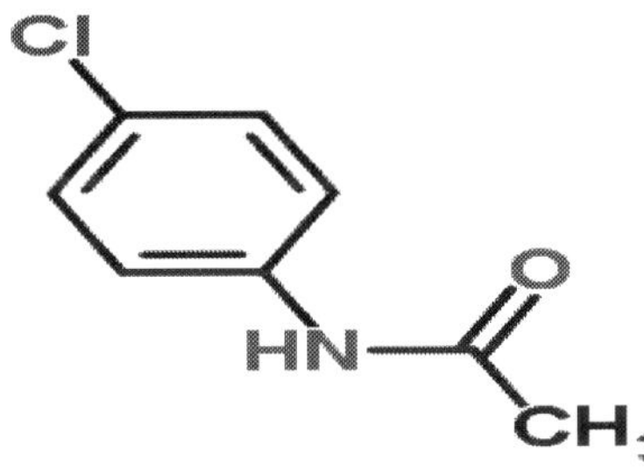

Figure 3.13: The chemical structure of chloroacetanilide derivatives.

Cyanazine (Figure 3.14) is a commonly used triazine with a water solubility more typical of the class (although at the high end): 171 ppm. It is stable to UV light, but hydrolyzed by pH less than 5 or greater than 9. It has a relatively low LD_{50} of 334 mg/kg, if given po to rats. The dermal LD_{50} in rabbits is 2,000 mg/kg which, although high enough not to be a major concern, is low for this class. High doses produce lethargy, inactivity, and depression in animals. There have been almost no reports of human toxicity due to cyanazine exposure. And, in contrast to many other triazines, cyanazine is negative in the whole battery of *in vitro* (cell-based) mutagenesis assays, as well as in the *in vivo* carcinogenesis assay run in mice. It is, however, a developmental toxicant, causing birth defects in rats at doses as low as 25 mg/kg/day, with a calculated NOAEL of 1 mg/kg. Fetal death occurs in rabbits at doses of 2 mg/kg/day. This developmental toxicity has led to restrictions on its use.

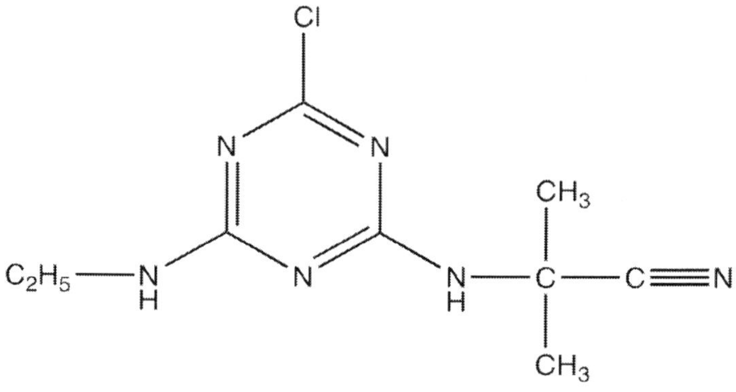

Figure 3.14: The chemical structure of cynazine.

Simazine (Figure 3.14) is the most persistent – and least water-soluble – of the commonly used triazines. It is soluble in water to 5 ppm, and has a 2-8 month persistence in Illinois soils. Like atrazine, it is more persistent in dry soils, but its persistence even under ordinary conditions of soil moisture is great enough that it is not used on corn because of the high probability that it would carry over into the next season, and damage the next season's soybeans(which are typically planted after corn). It is most often used as a soil sterilant, but also has crop uses. Until 1992 it was also used to control submerged aquatic weeds.

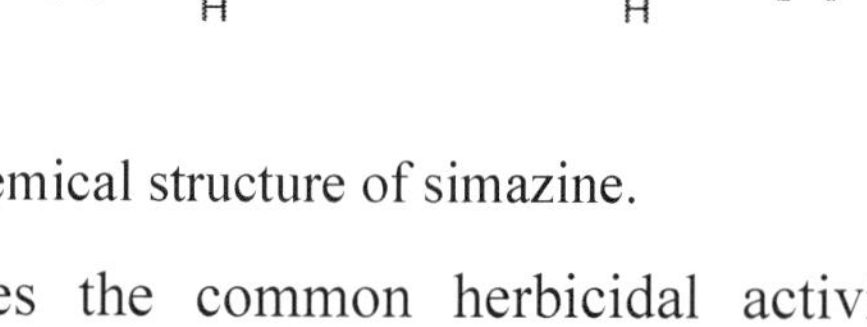

Figure 3.15: The chemical structure of simazine.

Simazine shares the common herbicidal activity of the class, interfering with the Hill reaction of photosynthesis. Its oral LD_{50} in rats is > 5,000 mg/kg; its dermal LD_{50} in rabbits is > 10,000 mg/kg. Sheep are somewhat more susceptible: the oral LD_{50} is approximately 500 mg/kg, and doses can be cumulative: e.g., a total dose of 1,400 mg/kg can be lethal. Symptoms of nonfatal simazine poisoning in mammals include tremors, convulsions, hindlimb weakness, decreased food intake and increased water intake. In chronic dosing, simazine increased the incidence of dominant lethals in *Drosophila melanogaster*, but was negative in all other mutagenesis assays. Similarly, it increased the incidence of mammary tumors in female rats, but was negative in the mouse assay. Therefore its mutagenicity and carcinogenicity are disputed. There is one study that suggests an increased prostate cancer risk among farmers exposed to simazine.

Studies on the Harmful Effects of Herbicides to Human Health

The harmful effects of herbicides on human health are determined by several factors, such as the chemical class of those compounds, dose, time, and exposure route. Herbicides can be toxic to humans at high and lower doses. The prolonged exposure can lead to a number of health effects, including the induction of diseases such as cancer and neurodegenerative, reproductive and developmental changes and respiratory effects.

Doll and Peto (1981) estimated that 35% of all cases of cancer in the U.S. population originate from diet, and the herbicides present in foods are responsible. Estrogenicity assays made by Hernández et al. (2012), show that organochlorine pesticides may act as endocrine disruption through more than one mechanism, including agonist or antagonist effects of different receptors. Chloro-s-triazize herbicides, pre-emergent pesticides used worldwide, have been generally considered as chemical compounds of low toxic potential for humans; however, there are many controversies on this issue. According to several international agencies, including the Environmental Protection Agency (EPA), Development for Environmental Assessment Center of the United States and IARC Monographs (International Agency for Research on Cancer), the herbicide atrazine, for example, was classified as a chemical agent probably carcinogenic to humans, although the basis for this conclusion is only evidenced in other animals. Due to the fact that atrazine induce mammary tumours in female Sprague-Dawley rats, the Peer Review Committee of the EPA Office of Pesticide Program (OPP) also concluded that atrazine should be considered in the Possibly Carcinogenic to Humans Group. However, EPA has classified this herbicide as a compound probably non carcinogenic to humans.

Some experimental studies have shown that exposure of humans to high doses of atrazine can result in an increased loss of body weight.

However, a great number of epidemiological studies carried out with workers occupationally exposed to triazine herbicides indicate that these compounds do not have carcinogenic potential for these individuals. By analyses of different studies, it was observed that, although the chloro-s-triazine herbicides interfere in the endocrine responses of different species of mammals, their potential impact on humans seem to be mainly related to reproduction and development and not with human carcinogenesis.

Gammon et al. (2005) discussed the extensive list of epidemiological studies with the herbicide atrazine, which describes that the carcinogenic potential of this compound to humans is not conclusive, although some studies have indicated a relationship between a high risk of prostate cancer and exposure to the herbicide.

Mladinic et al. (2009) evaluated the genotoxic and mutagenic effects of low concentrations of the herbicides glyphosate and terbuthylazine, considered safe and, therefore, considered possible to occur in occupational and residential exposures (ADI – Acceptable Daily Intake, REL – Residential Exposure Level, OEL – Occupational Exposure Level, and 1/100 and 1/16 LD_{50} – Lethal Dose 50% - oral, rat), in human lymphocytes, with and without the use of metabolic activation (S9 fraction), by the FSH cytome assay, using pan-centromeric DNA probes to assess the content of micronuclei and other chromatinic instabilities. The authors verified that the frequencies of micronuclei, nuclear buds and nucleoplasmic bridges of cells treated with glyphosate slightly increased after the concentration of OEL 3.5 μg/mL, but no concentration induced an increase of the centromeric signals (C+) or DAPI (DAPI+). Now, the treatment with the herbicide terbuthylazine without metabolic activation showed a dose-response increase in the frequency of micronuclei of the lymphocytes exposed, and the significant data were from the concentration of 0.0008 μg/mL (REL) tested. The concentrations ADI (0.00058 μg/mL), REL (0.0008 μg/mL) and OEL (0.008 μg/mL) of terbuthylazine induced a significant occurrence of micronuclei hybridized with the centromeric probe (C+), regardless the presence or absence of S9, and of nuclear buds containing centromeric signals, only in the presence of S9. By the results obtained, it was suggested that the lowest concentrations of glyphosate do not have relevant harmful effects for the DNA molecule, while terbuthylazine presents a predominant aneugenic potential for the genetic material of human lymphocytes.

Terbuthylazine belongs to the chloro-s-triazine herbicides class, which inhibits the photosynthesis of weeds, by reaching the photosystem II. It is a chemical used for a variety of crops, such as maize, sugarcane, olive and pineapple. Since the banishment of atrazine in European countries in 2006, terbuthylazine was recommended as its substitute. Due to the fact that the herbicide terbuthylazine is suspect of causing diseases in humans, such as non-Hodgkin lymphoma and lung cancer, Mladinic et al.(2012) evaluated the effects of prolonged exposure (14 days) to low concentrations of this compound (0.58 ng/ml and 8 ng/ml) in human lymphocytes, using the comct assay and the comet-FISH assay (with the c-Myc and TP 53 genes). Treatment with terbuthylazine induced the migration of fragments of DNA in a significant manner, only for the highest concentration treated. The results showed an impairment of the structural integrity of c-Myc and TP 53, due to the prolonged exposure of human lymphocytes to terbuthylazine. The fact that several copies of TP53 were affected by the herbicide can indicate its ability to negatively interfere in the control of the cell cycle. However, the authors concluded that, for a more detailed assessment of the risk of cancer associated with exposure to terbuthylazine, it should be evaluated the impact of this pesticide on other housekeeping genes and markers.

Mladinic et al. (2012) evaluated the genotoxic potential, by the comet assay and FISH, and oxidative damages, by the TBARS lipid peroxidation, of different concentrations of glyphosate (three similar to those observed in residential and occupational exposures and two related to LC_{50}) in human lymphocytes. The comet assay showed that the concentration of 580 μg/mL promoted a significant increase in the tail length, while the concentration of 92.8 μg/mL caused an increase in the tail intensity, both in relation to the control test. With the addition of the S9 fraction, the tail length was significantly increased for all the concentrations tested. When the lymphocytes were exposed to the three highest concentrations without S9, there was an increase in the frequency of micronuclei, nuclear buds and nucleoplasmic bridges. The addition of a metabolic activation system only promoted a significant increase of the nuclear instabilities for the highest concentration tested. The values of TBARS significantly increased with the increase of the concentrations tested, regardless the presence or absence of the S9 fraction. Due to the fact that dose-dependent effects for all the assays used were not observed, the authors concluded that these concentrations of glyphosate are not relevant for human exposure, since they did not present a significant risk for human health.

The study on the influence of a complex mixture of herbicides (atrazine, 2,4-D, alachlor, ciazine and malathion) in workers occupationally exposed to them, was carried out using cytogenetic methods standardly established (chromosome aberrations and micronucleus assay) and the comet assay technique. This assay showed a significant increase in the DNA migration ($P<0.001$), suggesting that long-term exposure to the pesticides could cause damages in the genome of somatic cells and, therefore, would represent a potential risk to human health.

EFFECTS OF FUNGICIDES ON HUMAN HEALTH

Fungicide Exposures

One of the more critical determinants of fungicide toxicity is the dose-time relationship. Dose is the quantity of the fungicide that something is exposed to while the time represents the frequency and duration of exposure. These dose-time relationships give rise to both acute and chronic toxicities. Acute toxicity refers to the toxicity of a fungicide after a short-term exposure. The effects of this type of toxicity are seen immediately following exposure. The acute toxicities of fungicides are the basis for label statements such as danger, warning, and caution. Chronic toxicities, on the other hand, refer to the delayed toxic effects from fungicide exposure. Those fungicides with high acute toxicity may not necessarily have high chronic toxicity. Similarly, those with low chronic toxicity may not necessarily have low acute toxicity One of the best ways to reduce the toxicity of a fungicide, whether it be acute or chronic, is simply to reduce your exposure.

Fungicide Toxicities

The toxicities offungicidesare usually determined as a function of either acute dermal, oral, or inhalation exposures. These toxicities are expressed either as LD50 or LC50. The LD50 is the lethal dose required to kill 50% of the test animals at a given time, whereas LC50 is the lethal concentration (in water or air) that kills 50% of the test animals at a given time. The following table indicates how certain toxicity levels are classified by the US Environmental Protection Agency (EPA) as shown in table 3.1.

Table 3.1: Categories of Acute Fungicide Toxicity

Toxicity Rating (Signal Words)	EPA Classification	Type of Exposure	Amount of Exposure	Probable Lethal Dose (150-lb/man)
Very High (Danger, Poison)	I	Oral Dermal Inhalation	0 -50 mg/kg 0-200 mg/kg 0 -0.2 mg/l	0 – 1 tsp.
High (Warning)	II	Oral Dermal Inhalation	50 -500 mg/kg 200-2,000 mg/kg 0.2- 2 mg/l	1 tsp. – 1 oz.
Medium (Caution)	III	Oral Dermal Inhalation	50 -500 mg/kg 200-2,000 mg/kg 0.2- 2 mg/l	1 oz. – 1 pint
Low (Caution)	IV	Oral Dermal Inhalation	Over 5,000 mg/kg over 20,000 mg/kg over 20 mg/l	

LD50 or LC50 of fungicides within each rating class.

EFFECTS OF MOLLUSCIDES ON HUMAN HEALTH

Molluscicides plays an important role in the effective way of eradicating schistosomiasis has been well established. Partly as a consequence of the rising costs of synthetic molluscicides and increase in investigation the different molluscicidal properties. An appealing aspect of natural plant products is that, while they may be highly toxic, they degrade very rapidly when released into the environment.

The role that molluscicides can play in the effective control of schistosomiasis (as shown in figure 3.14) has been well established.

Partly as a consequence of the rising costs of synthetic molluscicides, there has been an increasing interest in plants showing molluscicidal properties. An appealing aspect of natural plant products is that, while they may be highly toxic, they degrade very rapidly when released into the environment.

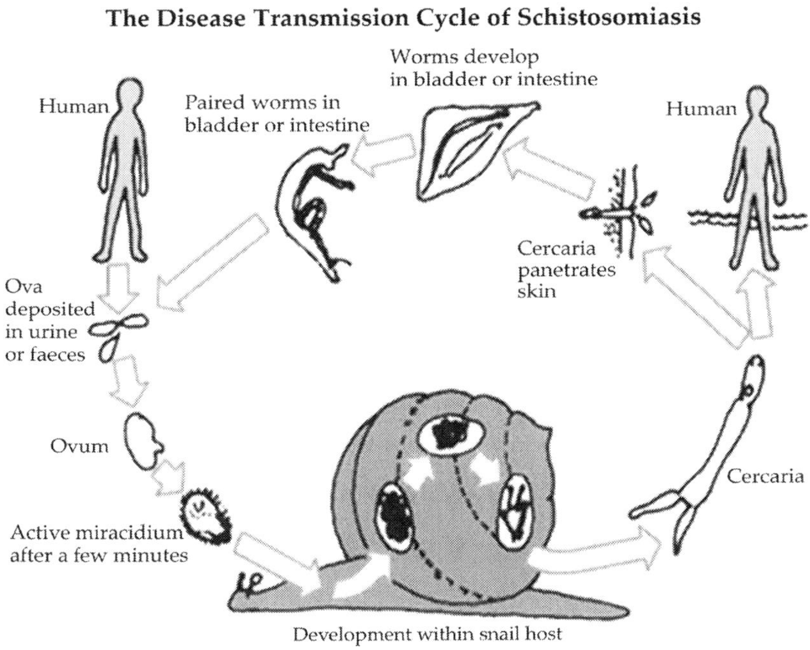

Figure 3.16: Transmission of a snail that causes schistosomiasis common to developing countries.

During the past 20 years, successful control of schistosomiasis has been achieved by a dosage of antibiotics using either in a single and combined methods of intervention and underlining the valuable role of synthetic molluscicides in the strategies applied. During the 1960's the use of molluscicides provided the only reliable approach to control of schistosomiasis, but with the advent of new and more efficient antischistosomal drugs during the last decade and to slowly kill the eggs in the intestines to removed the scolex free from the intestine through integrated methods have been increasingly used.

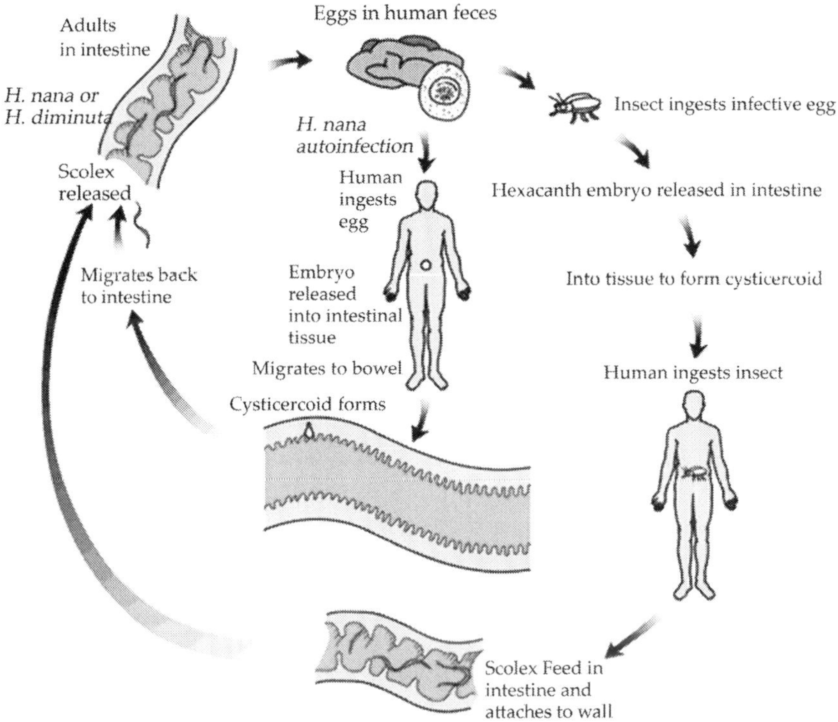

Figure 3.17: Transmission of molluscicides in human body.

Snail control by periodic area-wide applications of synthetic molluscicides 1s now being carried out successfully in Egypt in large irrigation schemes with controlled water management. "Transmission control", based upon the focality and seasonality of transmission in most situations, is also being successfully effected by molluscicidíng and surveillance. This may be a very cost-effective approach, if it can be related to accurate knowledge of human water-contact patterns, and if it can result in an important reduction of the need for expensive chemicals. Thus a reduction in transmission may be achieved by lowering snail population density and altering its age structure, even though complete eradication cannot be attained. Available data indicate that molluscicidíng may be cost-effective where the volume of water to be treated per person at risk is small.

A significant reduction of the disease may be possible using the current technology and employing the new efficient drugs. Chemotherapy will undoubtedly play a vital role in reducing the severity of the disease and contribute significantly to the control of transmission by lowering egg output. However, the effect on transmission of population-based chemotherapy as a single control measure is not known. Nevertheless in many endemic areas, integrated measures involving snail control and chemotherapy will reduce the prevalence and intensity of infection and control transmission, cost-effectively, within a short period of time when schistosomiasis is not given high priority, however, the cost of drugs, molluscicides and other methods may deter the implementation of any control operations.

It 1s doubtful that the present generation with the exception of niclosamide, will continue a serious problem for future snail control. of synthetic molluscicides, to be available. This poses

Thus, economic and ecological considerations increasingly favour the use of molluscicides with selective activity, which are biodegradable, inexpensive and readily available in endemic areas. The impetus to study possible plant molluscicides is due largely to the high cost of imported synthetic compounds and the desirability of introducing appropriate technology into endemic areas.

REFERENCES

1. Abdollahi M, Ranjbar A, Shadnia S, Nikfar S, Rezaie A. Pesticides and oxidative stress: a review. Medical Science Monitor 2004;10:141-147.
2. Agriculture and Natural Resources. Water quality: Controlling Nonpoint Source (NPS) Pollution. Pesticide Management To Protect Water Quality. Understanding Pesticides And How They Affect Water Quality 1999. Available from http://www.aces.edu/pubs/docs/A/ANR-0790/WQ4.5.1.pdf (accessed 2 September 2014).
3. Akcha F, Spagnol C, Rouxel J. Genotoxicity of diuron and glyphosate in oyster spermatozoa and embryos. Aquatic Toxicology 2012; 106-107: 104-113.
4. Albaseer SS, Nageswara Rao R, Swamy YV, Mukkanti K.

Analytical artifacts, sample handling and preservation methods of environmental samples of synthetic pyrethroids. Trends in Analytical Chemistry 2011; 30 [11]: 1771-1780.

5. Almeida EA, Bainyb ACD, Dafrec AL, Gomesa OF, Medeirosa MHG, di Mascioa P. Oxidative stress in digestive gland and gill of the brown mussel Perna perna exposed to air and re-submersed. Journal of Experimental Marine Biology and Ecology. 2005; 318: 21–30.
6. Alonzo HGA, Corrêa CL. Praguicidas. In: Oga S, Camargo MMA, Batistuzzo JA. (eds) Fundamentos de Toxicologia. 4ª. Ed. Editora Atheneu, São Paulo; 2014. p.323-341.
7. Anadón A, Martínez-Larrañaga MR, Martínez MA. Use and abuse of pyrethrins and synthetic pyrethroids in veterinary medicine. The Veterinary Journal 2009; 182: 7-20.
8. Andreu V, Picó Y. Determination of currently used pesticides in biota. Analytical and Bioanalytical Chemistry 2012; 404[9]:2659-81.
9. Ashby J, Kier L, Wilson AG, Green T, Lefevre PA, Tinwell H, Willis GA, Heydens WF, Clapp MJ. Evaluation of the potential carcinogenicity and genetic toxicity to humans of the herbicide acetochlor. Human & Experimental Toxicology 1996; 15: 702-735.Heydens WF, Lamb IC, Wilson AGE. Chloroacetanilides. In: Krieger R (ed). Handbook of Pesticide Toxicology. San Diego, CA: Academic Press; 2001. p. 1543–1558.
10. Bateman DN. Pharmacological treatments of paraquat poisoning. Human Toxicology 1987; 6[1]: 57-62.
11. Bedos C, Cellier P, Calvet R, Barriuso E, Gabrielle B. Mass transfer of pesticides into the atmosphere by volatilization from soils and plants: overview. Agronomie 2002; 22: 21–33.
12. Bergeson LL. Nanosilver: US EPA's pesticide office considers how best to proceed. Environmental Quallity Management 2010; 19 [3]: 79−85. http://onlinelibrary.wiley.com/doi/10.1002/tqem.20255/pdf (Accessed 2 September 2014).
13. Bergman A, Heindel JJ, Jobling S, Kidd KA, Zoeller T. State of the Science of Endocrine Disrupting Chemicals – 2012. United Nations Environment Programme and the World Health Organization 2013.

http://www.who.int/ceh/publications/endocrine/en (Accessed 3 September 2014).

14. Bjorling-Poulse M, Andersen HR, Grandjean P. Potential developmental neurotoxicity of pesticides used in Europe. Environmental Health. 2008; 7-50.
15. Bradberry SM, Cage SA, Proudfoot AT, Vale JA. Poisoning due to Pyrethroids. Toxicology Reviews 2005; 24 [2]: 93-106.
16. Bradberry SM, Proudfoot AT, Vale A. Glyphosate Poisoning. Toxicology Reviews 2004; 23: 159-167.
17. Cai Z, Wang D, Ma WT. Gas chromatography/ion trap mass spectrometry applied for the analysis of triazine herbicides in environmental waters by an isotope dilution technique. Analytica Chimica Acta 2004; 503: 263–270.
18. Campbell AW. Glyphosate: its effects on humans. Alternatives Therapies in Health and Medicine 2014; 20: 9-11.
19. ÇAVAS TC, KONEN S. Detection of cytogenetic and DNA damage in peripheral erythrocytes of goldfish (Carassius auratus) exposed to a glyphosate formulation using the micronucleus test and the comet assay. Mutagenesis 2007; 22: 263-268.
20. Chan PC, Mahler JF. NTP technical report on the toxicity studies of glyphosate (CAS no. 1071-83-6) administered in dosed feed to F344/N rats and B6C3F1 mice. Toxicity Reports Series 1992; 16: 1-D3.
21. Costa LG. Toxic Effects of Pesticides. In: Klaassen CD. (ed). Cassarett and Doull's Toxicology *The Basic Science of Poisons*. 7 Ed, McGraw-Hill, New York, 2008. p.883-930.
22. Czaja K, Góralczyk K, Struciński P, Hernik A, Korcz W, Minorczyk M, Lyczewska M, Ludwicki JK. Biopesticides - towards increased consumer safety in the European Union. Pest Manag Sci. 2014; In press.
23. Davies JE, Barquet A, Freed VH, Haque R, Morgade C, Sonneborn RE, Vaclavek C. Human pesticide poisonings by a fat soluble organophosphate insecticide. Archives of Environmental Health 1975; 30: 608-613.
24. Delaplane K S. Pesticide Usage in the United States: History,

Benefits, Risks, and Trends. The University of Georgia, Athens, Georgia 1996. http://ipm.ncsu.edu/safety/factsheets/pestuse.pdf (Accessed 20 August 2014).

25. Di Stefano V, Avellone G, Bongiorno D, Cunsolo V, Muccilli V, Sforza S, Dossena A, Drahos L, Vékey K. Applications of liquid chromatography-mass spectrometry for food analysis. Journal of Chromatography A 2012; 1259: 74-85.
26. Durmaz H, Sevgiler Y, Üner N. Tissue-speciWc antioxidative and neurotoxic responses to diazinon in Oreochromis niloticus. Pesticide Biochemistry and Physiology 2006; 84: 215–226.
27. Ecobichon DJ. Carbamate insecticides. In: Krieger R (ed). Handbook of Pesticide Toxicology. San Diego, CA: Academic Press; 2001. p.1087-1106.
28. Ellenhorn MJ, Barceloux DG. Medical Toxicology. Diagnosis and Treatment of Human Poisoning. Ed. Elsevier, New York, 1988. 1512p.
29. EPA - Environmental Protection Agency. Overview of the Ecological Risk Assessment Processin the Office of Pesticide Programs,U.S., Environmental Protection Agency. 2004. http://www.epa.gov/espp/consultation/ecorisk-overview.pdf (Acessed 13 August 2014).
30. Ermler S, Scholze M, Kortenkamp A. Seven benzimidazole pesticides combined at sub-threshold levels induce micronuclei in vitro. Mutagenesis 2013; 28: 417-426.
31. Farenhorst A, Saiyed IM, Goh TB, McQueen P. The important characteristics of soil organic matter affecting 2,4-dichlorophenoxyacetic acid sorption along a catenary sequence. Journal of Environmental Science and Health, Part B. 2010; 45:204-13.
32. Fenik J, Tankiewicz M, Biziuk M. Properties and determination of pesticides in fruits and vegetables. Trends in Analytical Chemistry 2011; 30 [6]: 814-826.
33. Fenoll J, Vela N, Garrido I, Pérez-Lucas G, Navarro S. Abatement of spinosad and indoxacarb residues in pure water by photocatalytic treatment using binary and ternary oxides of Zn and Ti. Environmental science and pollution research international. 2014; *In press.*
34. Fortin MC, Bouchard M, Carrier G, Dumas P. Biological monitoring

of exposure to pyrethrins and pyrethroids in a metropolitan population of the Province of Quebec, Canada. Environmental Research 2008; 107: 343-350.

35. Freire C, Koifman S. Pesticides, depression and suicide: a systematic review of the epidemiological evidence. International Journal of Hygiene and Environmental Health 2013; 216[4]:445-460.
36. Gallo MA, Lawryk NJ. Organophosphorus pesticides. In: Hayes WJ Jr.; Laws ER Jr. (Eds) Handbook of Pesticide Toxicology. San Diego: Academic Press. 1991. 2: 917-1123.
37. Hai DQ, Varga SI, Matkovics B. Effects of diethyl-dithiocarbamate on antioxidant system in carp tissue. Acta Biologica Hungarica 1997; 48: 1-8.
38. Halliwell B, Gutteridge JMC. Free Radicals in Biology and Medicine, 3ª ed. Oxford University Press, Oxford. 1999.Lushchak VI. Environmentally induced oxidative stress in aquatic animals. Aquatic Toxicology 2011; 101: 13-30.
39. Hernke MT, Podein RJ. Sustainability, health and precautionary perspectives on lawn pesticides, and alternatives. Ecohealth. 2011; 8:223-32.
40. Ho IK, Fernando JC, Sivam SP, Hoskins B. Roles of Dopamine and GABA in neurotoxicity of organophosphorous cholinesterase inhibitors. Proceedings of the Western Pharmacoly Society 1984; 27: 177-180.
41. Honore P, Hantson P, Fauville JP, Peeters, A. Manieu, P. Paraquat poisoning: State of the art. Acta Clinica Belgica 1994; 49:220-8.
42. Hori Y, Tanaka T, Fujisawa M, Shimada K. Toxicokinetics of DL-glufosinate enantiomer in human BASTA poisoning. Biological and Pharmaceutical Bulletin 2003; 26: 540-543.
43. Hour BT, Belen C, Zar T, Lien YH. Herbicide roundup intoxication: successful treatment with continuous renal replacement therapy. American Journal of Medicine 2012; 125[8]: e1-2.
44. Inoue Y, Onodera M, Fujita T, Fujino Y, Kikuchi S, Endo S. Factors associated with severe effects following acute glufosinate poisoning. Clinical Toxicology 2013; 51: 846-849.
45. Jarrard HE; Delaney KR; Kennedy CJ. Impacts of carbamate

pesticides on olfactory neurophysiology and cholinesterase activity in coho salmon (Oncorhynchus kisutch). Aquatic Toxicology 2004; 69: 133-148.

46. Jokanovic M. Medical treatment of acute poisoning with organophosphorus and carbamate pesticides, Toxicology Letters 2009; 190: 107-115.

47. Jokanovic M. The Impact of Pesticides. 1a Ed. The Academy Publish, Cheyenne. 2012. 417p.

48. Kelly SA, Havrilla CM, Brady TC, Abramo KH, Levin ED. Oxidative Stress in Toxicology: Established Mammalian and Emerging Piscine Model Systems. Research Reviews 1998; 106: 375-384.

49. Kitamura S, Sugihara K, Fujimoto N. Endocrine Disruption by Organophosphate and Carbamate Pesticides. In: Gupta RC (ed.) Toxicoly of Organophosphonate and Carbamate Compounds. Elsevier Academic Press, New York, 2006. p. 481-194.

50. Knutson RD, Taylor CR, Penson JB, Smith EG. Economic impacts of reduced chemical use. Knutson and Associates, College Station, Texas. 1990.

51. Kookana RS, Boxal AB, Reeves PT, Ashauer R, Chaudhry Q, Cornelis G, Fernandes TF, Gan J, Kah M, Lynch I, Ranville J, Sinclair C, Spurgeon D, Tiede K, Van Den Brink PJ. Nanopesticides: guiding principles for regulatory evaluation of environmental risks. Journal of Agriculture and Food Chemistry 2014; 62: 4227-4240.

52. Levengood JM, Beasley VR. Principles of ecotoxicology. In: Gupta RC (ed.) Veterinary toxicology: basic and clinical principles. Academic press, Amterdan. 2007. p. 689-708.

53. Llasera MPG, González MB. Presence of carbamate pesticides in Environmental waters from the northwest of mexico: determination by liquid chromatography. Water Research 2001; 35 [8]: 1933-1940.

54. Marrs TC, Maynard RL. Neurotransmission system as targets for toxicants: a review. Cell Biology and Toxicology 2013; 29: 381-396.

55. Martins JG, Amaya Chávez A, Waliszewski SM, Colín Cruz A, García Fabila MM. Extraction and clean-up methods for organochlorine pesticides determination in milk. Chemosphere 2013; 92[3]: 233-46.

56. Matthews G, Bateman R, Miller P. Book Pesticide application methods. John Wiley & Sons. 2008. Available from http://onlinelibrary.wiley.com/book/10.1002/9780470760130 (Accessed 4 September 2014).

57. Mayhew DA, Taylor GD, Smith SH, Banas DA. Twenty-four month combined chronic oral toxicity and oncogencity study in rats utilizing atrazine technical grade. Lab Study No.: 410-1102, Accession No. 262714-262727, American Biogenics Corp., Decatur, 1986. p. 2-6.

58. Milatovic D, Gupta RC, Aschner M. Anticholinesterase toxicity, oxidative stress. Science World Journal 2006; 6:295-310.

59. Padhye LP, Kim JH, Huang CH. Oxidation of dithiocarbamates to yield N-nitrosamines by water disinfection oxidants. Water research 2013; 47: 725-736.

60. Park JS, Kwak SJ, Gil HW, Kim SY, Hong SY. Glufosinate Herbicide Intoxication Causing Unconsciousness, Convulsion, and 6th Cranial Nerve Palsy. Journal of Korean Medical Science 2013; 28: 1687-1689.

61. Paudyal BP. Organophosphorus poisoning. Journal of Nepal Medical Association 2008; 47(172): 251-258.

62. Ratcliffe D. Decreases in eggshell weight in certain birds of prey. Nature 1967; 215: 208-210.

63. Ray DE, Fry JR. A reassessment of the neurotoxicity of pyrethroid insecticides. Pharmacology & Therapeutics 2006; 111[1]: 174-193.

64. Reigart JR, Roberts JR. Recognition and Management of Pesticide Poisonings. 5ª Ed. United States Environmental Protection Agency, 1999. http://www.epa.gov/oppfead1/safety/healthcare/handbook/handbook.pdf (Accessed 4 September 2014).

65. Roberts T R, Dyson JS, Lane MCG. Deactivation of the Biological Activity of Paraquat in the Soil Environment: a Review of Long-Term Environmental Fate. Journal of Agriculture and Food Chemestry 2002; 50: 3623-3631.

66. Tien CJ, Lin MC, Chiu WH, Chen CS. Biodegradation of carbamate pesticides by natural river biofilms in different seasons and their effects on biofilm community structure. Environmental Pollution 2013; 179: 95-104.Li Y, Niu J, Shen J, Zhang C, Wang Z, He T. Spatial and seasonal distribution of organochlorine pesticides in the

sediments of the Yangtze Estuary. Chemosphere 2014; 114: 233-240.

67. Toygar M, Aydin I, Agilli M, Aydin FN, Oztosun M, Gul H, Macit E, Karslioglu Y, Topal T, Uysal B, Honca M. The relation between oxidative stress, inflammation, and neopterin in the paraquat-induced lung toxicity. Human and Experimental Toxicity, 2014. *In press*. Ver volume ou colocar link.

68. Unsworth 2010 - History of pesticide use. International Union of pure and applied chemistry (IUPAC). 2010. http://agrochemicals.iupac.org/index.php?option=com_ sobi2&sobi2Task =sobi2Details &catid=3&sobi2Id=31 (Accessed 2 September 2014).

69. USEPA - United States of Environmental Protection Agengy. About pesticides. U.S. EPA. http://www.epa.gov/pesticides/about/index.htm (Accessed 27 August 2014).

70. USEPA - United States of Environmental Protection Agengy. Special Report on Environmental Endocrine Disruption: An Effects Assessment and Analisys, U.S. Environmental Protection Agency, Report No. EPA/630/R-96/012, Washington D. C, 1997. http://www.epa.gov/raf/publications/pdfs/ENDOCRINE.PDF (Accessed 27 August 2014).

71. Valavanidis A, Vlahogianni T, Dassenakis M, Scoullos M. Molecular biomarkers of oxidative stress in aquatic organisms in relation to toxic environmental pollutants. Ecotoxicology and Environmental Safety 2006; 64: 178-189.

72. Van Dyk JS, Pletschke B. Review on the use of enzymes for the detection of organochlorine, organophosphate and carbamate pesticides in the environment. Chemosphere 2011; 82: 291-307.

73. Wallace KB. Target organ toxicology series. Taylor & Francis Ltd, Washington 442p. 1997.

74. Wen Y, Fu Z, Xu J, Tang S, Wang Q, Li H, Xie G, Zhu Y, Gu Y, Tan F. Determination of 2, 4-dichlorophenoxyacetic acid in air of workplace by high-performance liquid chromatography. Zhonghua Lao Dong Wei Sheng Zhi Ye Bing Za Zhi 2014; 32[6]: 458-459.

75. WHO-World Health Organization. Suicide prevention (SUPRE). http://www.who.int/mental _health/prevention /suicide/suicideprevent/en/ (Accessed 16 August 2014).

76. Williams GM, Kroes R, Munro IC. Safety evaluation and risk assessment of the herbicide Roundup and its active ingredient, glyphosate, for humans. Regulatory Toxicology and Pharmacology 2000; 31: 117-65.
77. Zhang Q, Zhang W, Wang X, Li P. Immunoassay Development for the Class-Specific Assay for Types I and II Pyrethroid Insecticides in Water Samples. Molecules 2010; 15: 164-177.
78. Jurado, A.S., Fernandes, M.A.S., Videira, R.A., Peixoto, F.P., Vicente, J.A.F. Herbicides: The Face and the Reverse of the Coin. An *in vitro* Approach to the Toxicity of Herbicides in Non-Target Organisms. In: KORTEKAMP, A. (Ed.) Herbicides and Environment. Kroatia, 2011, p. 3- 45 p.
79. Moreland, D.E. Mechanisms of action of herbicides. Ann. Rev. Plant Physiol., v. 31, p. 597-638, 1980.
80. Rao, V.S. Principles of Weed Science, New Hampshire, USA, 2nd Ed, 2000, 559 p.
81. Zeliger, H.I., Human toxicology of chemical mixtures. In: Toxic Consequences Beyond the Impact of One-component Product and Environmental Exposures. 2nd ed. Elsevier, Oxford, 2011.
82. Bassil, K.L.; VAKIL, C.; SANBORN, M.; COLE, D.C.; KAUR, J.S.; KERR, K.J. Cancer health 585 effects of pesticides: systematic review. Can. Fam. Physician, v. 53, p. 1704-1711, 2007.
83. Parrón, T.; Requena, M.; Hernández, A.F.; Alarcón, R. Association between environmental exposure to pesticides and neurodegenerative diseases. Toxicol. Appl. Pharmacol., v. 256, p. 379-385, 2011.
84. Hanke, W.; Jurewicz, J., The risk of adverse reproductive and developmental disorders due to occupational pesticide exposure: an overview of current epidemiological evidence. Int. J. Occup. Med. Environ. Health, v. 17, p. 223-243, 2004.
85. Hernández, A.F.; Parrón, T.; Alarcón, R. Pesticides and asthma. Curr. Opin. Allergy Clin. Immunol., v. 11, p. 90-96, 2011.
86. Doll, R.; Peto, R. The causes of cancer: quantitative estimates of avoidable risks of cancer in the United States today. Journal National Cancer Institute, v. 66, p. 1191-1308, 1981.
87. Hernández, A.F.; Parrón, T.; Tsatsakis, A.M.; Requena, M.;

Alarcón; López-Guarnido, O. Toxic effects of pesticide mixtures at a molecular level: Their relevance to human health. Toxicology, 2012 (*in press*).

88. Jones, F.; Fawell, J.K. Lessons learnt from the river DEE Pollution Incident. Public Health in Proceedings of the Word conference on chemicals accidents, Harvard, v. 4, p. 223-226, 1987.
89. Waters, M.D.; Stack, H.F.; Jackson, M.A. Genetic toxicology data in the evaluation of potential human environmental carcinogens. Mutation Research, Amsterdam, v. 437, p. 21-49, 1999.
90. USEPA. Atrazine: Carcinogenicity characterization and hazard assessment, office of pesticide programs, health effects division. 1999. Http:/ www.epa.gov/scipoly/sap/#jan
91. EPA, 2003. US Environmental Protection Agency. October 31, 2003, revised atrazine interim reregistration eligibility decision (IRED). Office of Prevention, Pesticides and Toxic Substances, 2003.
92. Jowa, L.; HOWD, R. Should atrazine and related chlorotriazines be considered carcinogenic for human health risk assessment? J Environ Sci Health C Environ Carcinog Ecotoxicol Rev., v. 29 (2), p. 91-144, 2011.
93. Mladinic, M.; Perkovic, P.; Zeljezic, D. Characterization of chromatin instabilities induced by glyphosate, terbuthylazine and carbofuran using cytome FISH assay. Toxicology Letters, v. 189, p. 130-137, 2009b.
94. Gebel, T.; Kevekordes, S.; Pav, K.; Edenharder, R.; Dunkelberg, H. *In vivo* genotoxicity of selected herbicides in the mouse bone-marrow micronucleus test. Arch. Toxicol., v. 71, 193-197, 1997.
95. Mladinic, M.; Zeljezic, D.; Shaposhnikov, S.A.; Collins, A.R. The use of FISH-comet to detect c-Myc and TP 53 damage in extended-term lymphocyte cultures treated with terbuthylazine and carbofuran. Toxicology Letters, v. 211, p. 62-69, 2012.
96. Hardell, L.; Eriksson, M.; Nordstrom, M. Exposure to pesticides as risk factor for non-Hodgkin's lymphoma and hairy cell leukemia: Pooled analysis of two Swedish case-control studies. Leuk Lymp, v. 43, p. 1043-1049, 2002.
97. De Roos, A.J.; Zahm, S.H.; Cantor, K.P.; Weisenburger, D.D.; Holmes, F.F.; Burmeister, L.F.; Blair, A. Integrative assessment of

multiple pesticides as risk factors for non-Hodgkin's lymphoma among men. Occup Environ Med, v. 60, p.11, 2003.

98. He, X.; Wang, L.; Szklarz, G.; Bi, Y.; Ma, Q. Resveratrol Inhibits Paraquat-Induced Oxidative Stress and Fibrogenic Response by Activating the Nuclear Factor Erythroid 2-Related Factor 2 Pathway. The Journal of Pharmacology and Experimental Therapeutics, v. 342, p. 81-90, 2012b.

99. Ossowska, K; Smiałowska, M.; Kuter, K.; Wieron´ Ska, J.; Zieba, B.; Wardas, J.; Nowak, P.; DABROWSKA, J.; BORTEL, A.; BIEDKA, I. et al. Degeneration of dopaminergic mesocortical neurons and activation of compensatory processes induced by a long-term paraquat administration in rats: implications for Parkinson's disease. Neuroscience, v. 141, p. 2155-2165, 2006.

100. Tanner, C.M.; Kamel, F.; Ross, G.W.; Hoppin, J.A.; Goldman, S.M.; Korell, M.; Marras, C.; Bhudhikanok, G.S.; Kasten, M.; Chade, A.R. et al. Rotenone, paraquat, and Parkinson's disease. Environ Health Perspect, v. 119, p. 866-872, 2011.

4
CHAPTER

PESTICIDES ARE POISON

INTRODUCTION

A **pesticide** are compounds that used to eradicate animals, insects, fungi or plants that is not useful to mankind.Actually there are thousands of different pesticides that are used in residents, commercial shops, offices, storerooms, sheds, gardens, farms, pastoral stations and many other places.Most of the pesticides used today are chemicals which have been developed in a laboratory by scientists and produced in factories.

Some pesticides are quite hazardous, as they can be harmful to humans and other living things.Due to the presence of chemical residuethey tend to contaminate land, the air, food crops, water ways and seriously harm or kill native animals, pets and domestic animals. In addition to being hazardous to the user, pesticides can also cause great harm and sometimes death to a person or other living things nearby, if the instructions on the pesticide container are not followed carefully.

Many insects are pests eats and consume many of the food plants farmers produce and farmers tried some techniques to stop this insects infestations. Pesticides are the latest human invention for killing pest insects (the '-icide' bit of 'pesticide' means 'killer'). Pesticides are poisonous chemicals which farmers spray on their crops. The poison kills the insects. End of problem. Well not quite. The pesticides don't just disappear. They stick around and poison many other creatures (including humans - over 200,000 people are thought to die every year because of pesticide poisonings).

Untold stories of innocent animals and huma being accidentally due to it's potency. Pesticides also pollute water supplies and plants that have been sprayed with them often contain small amounts when people eat them. People who support genetic engineering claim that if farmers use GE plants which make their own pesticides, then farmers will need to spray less pesticide which will be better for everyone.

PESTICIDES CAUSE MANY HEALTH PROBLEMS

Pesticides are hazardous compounds released intentionally into our environment to kill living organisms This includes substances that kill weeds (herbicides), insects (insecticides), fungus (fungicides), rodents (rodenticides), and others.

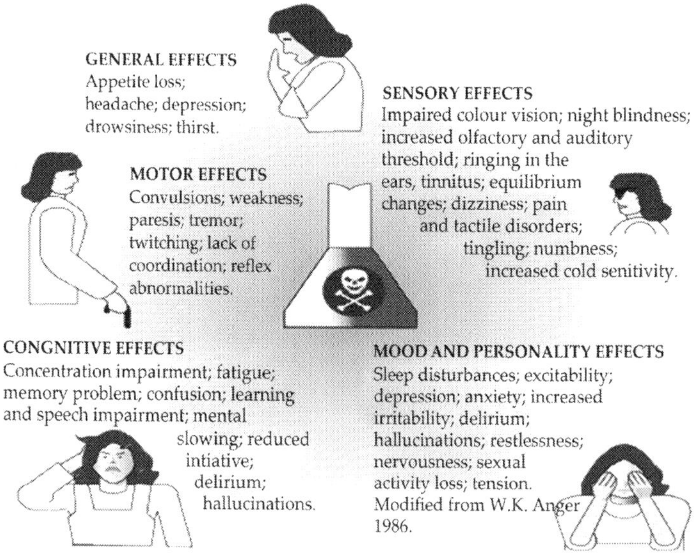

Figure 4.1: Neurological and behavioural effects of exposure to neurotoxic chemicals such as pesticides.

The application of pesticides to manage pest problems has become a common practice in agriculture, industry and even in homes around the world. Pesticides are used almost everywhere which is not only in agricultural fields, but also in homes, parks, schools, buildings, forests, and roads. It is difficult to find somewhere where pesticides aren't used -- from the can of bug spray under the kitchen sink to the airplane crop dusting acres of farmland, our world is filled with pesticides. In addition, pesticides can be found in the air we breathe, the food we eat, and the water we drink.

Most studies of the health effects of pesticides have focused on occupationally exposed people, like farmworkers and pesticide applicators. Acute OP pesticide poisonings result in symptoms like nausea, abdominal cramps, diarrhea, dizziness, anxiety and confusion, which can be quite severe but are often reversible. Figure 5.1 shows the neurological effect of pesticides to the brain of human beings and table 5.1 shows the effect of pesticde to human body due the presence of toxic compounds.

Table 4.1: Effects of Pesticide Chemicals on the Human Body Chemical Family Type of Pesticide Action on Human

Effects of Pesticidde Chemicals on the Human Body

Chemical Family	Type of Pesticide	Action on Human System	Symptoms Internal Exposure	Symptoms External Exposure	Symptoms Chronic Exposure
Organo-phosphates diazinon^{C} Spectracide^{T}	Insecticides, acaricides	Inhibits acetylcholi-nesterase (an enzyme) in the tissues	Headache, dizziness, weak-ness, shaking, nausea, stomach cramps, diarrhea, sweating	Minimal rashes but readily ab-sorbed through the skin	Loss of appetite, weakeness, weight loss, and general feeling of sickness
Carbamates carbaryl^{C} Sevin^{T}	Insecticides acaricides	Reversible changes in acetylcholines-terase enzyme of tissues	Headache, dizziness, weak-ness, shaking, nausea, stomach cramps, diarrhea, sweating	Minimal rashes but readily ab-sorbed through the skin	Loss of appetite, weakeness, weight loss, and general feeling of sickness
Organo-chlorines (Chlorinated Hydrocarbons methoxychlor^{C} Marlate^{T})	Insecticides, acaricides (HCB is a fungicide)	Disrupt function of nervous system, mainly the brain	Headache, dizziness, weak-ness, shaking, nausea, excita-bility, disor-ientation	Minimal rashes but readily ab-sorbed through the skin	Some bulldup in the fat tissues. May cause nerv-ousness, weak-ness, and shaking

There have also been many studies in groups of people who work with pesticides but who have not experienced acute poisonings serious enough to result in these kinds of symptoms. These studies have found that chronic, lower dose exposure is associated with respiratory problems, memory disorders, skin conditions, depression, miscarriage, birth defects, cancer and neurological conditions such as Parkinson's disease. There have been fewer studies of people without known occupational exposures, but one study

with a nationally representative sample showed increasing odds of ADD/ADHD for 8-15 year olds with increasing levels of OP pesticides metabolites in urine.

On May 2010, scientists from the University of Montreal and Harvard University released a study found out that exposure to pesticide residues on vegetables and fruit may double a child's risk of attention deficit hyperactivity disorder (ADHD), this is a condition that can cause inattention, hyperactivity, and impulsivity in children.

Fetuses, infants, growing children, pregnant and nursing mothers, and women of childbearing age are most at risk for adverse health outcomes from exposure to pesticides. Children are more at risk than adults because children eat more relative to their body weight than adults eat. Exposures during vulnerable periods of development can be particularly dangerous. These vulnerable period include fetal development, infancy, early childhood, and puberty. Fetuses are exposed to pesticides through the mother's diet. Infants are exposed through breast milk.

Acute poisoning is a persistent problem among agricultural and occupational workers who handle pesticides and track them into their homes where family members get exposed. People who live near agricultural fields may be exposed by drift from aerial spraying. Pesticides are also used in our schools, parks, and public lands. Pesticides are sprayed on agricultural fields and wood lots. Pesticides can be found in our air, our food, our soil, our water and even in our breast milk.

CHILDREN AND PESTICIDE POISONING

According toLewis et al. (2006), that children encounter pesticides daily in air, food, dust, and soil and on surfaces through home and public lawn or garden application, household insecticide use, application to pets, and agricultural product residues. For many children, diet may be the most influential source, as illustrated by an intervention study that placed children on an organic diet (produced without pesticide) and observed drastic and immediate decrease in urinary excretion of pesticide metabolites (Lu, Toepel, Irish, Fenske, Barr& Bravo,2006).

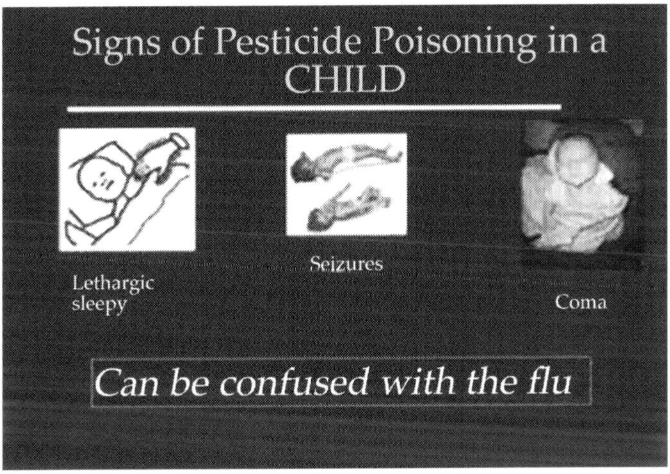

Figure 4.2: Children affected by pesticide residues in air, water and soil.

In agricultural settings, pesticide spray drift is important for residences near treated crops or by take-home exposure on clothing and footwear of agricultural workers (Shipp, Cooper, del Junco, Bolin&Whitworth, 2007). Teen workers may have occupational exposures on the farm or in lawn care (Eckerman, Gimenes,de Souza , Lopes ,Galvão, Sarcinelli & Chrisman, 2007). Heavy use of pesticides may also occur in urban pest control (Landrigan,Claudio &Markowitz, 1999).

National population-based surveys of pesticide urinary metabolites in adults indicate widespread exposure to pesticides, including organophosphates, carbamates, wood preservatives, and fungicides (Hill, Head, Baker, Gregg , Shealy , Bailey , Williams , Sampson & Needham, 1995). For example, Hill et al. (1995) detected chlorpyrifos an organophosphate pesticide, in 82% of 993 adults tested through the National Health and Nutrition Examination Survey, and found a 5-fold increase in the proportion of adults with levels over 5 pg/L compared to earlier surveys, suggesting increasing exposure in the general population.

OPs are eliminated from the body after 3-6 days (Bradway , Shafik & Lores ,1997), so the widespread detection of these compounds indicates continuing exposure.

Biologic information on children's pesticide exposure is very limited. Hill et al. (1995) reported detections ofdichlorobenzene and wood preservatives in 96% and 100%, respectively, of 197 Arkansas children, whereas phenoxy herbicide metabolites were found in 20% of all samples.Preliminary results from the federal Agricultural Health Study indicate detectable pesticide residues in children's urine (Steen, Bond & Mage, 1997). In the investigation made by Loewenherz, Fenske, Simcox, Bellamy & Kalman (1997) working in Washington state, found that 44% ofchildren of pesticide applicators and 27% of nonfarm, rural children had detectable OP residues.

In preliminary data from Arizona, chlorpyrifos was detected in 100% of about 40 children >6 years of age sampled in a population-based survey, and approximately 25% of 150 children <6 years of age sampled in an agricultural area. Detection limits in the second survey were higher (O'Rourke, 1998). Comparison of these studies to each other and to data reported by Hill et al. (1995) is difficult because of differences in detection limits, sample type (spot samples vs first morning void), and ages of participants. Overall, these studies suggest the potential for widespread low-level pesticide exposure in children and the need for population-based studies to establish norms.

Pesticides enter children's bodies via dermal absorption, ingestion, and inhalation. Exposure in the home depends on the frequency, duration, and nature such as skin contact or either by ingestion through hand to mouth transfer of the child's interaction with contaminated media such as house dust. Children may have higher exposure to pesticides than other residents living in the same contaminated environment, in part because young children spend more of their time indoors at home than adult (Silvers, Florence, Rourke& Lorimor,1994; CARB,1999). Most of them are likely to spend more time near the sorrounding in proximity to any pesticides present in their immediate environment. The importance of specific exposure-related behaviors, such as hand to-mouth activity, will be age dependent, suggesting that consequent exposure levels and pathways will vary with age, as has been observed for lead exposure (Clark, Bornschein,Succop,Que Hee, Hammond & Peace,1985).

For example, children younger than 6 months of age may receive their greatest exposures through breast milk or inhalation, but dermal absorption and ingestion may be the major pathway of exposure when children begin crawling and placing their hands on dusty surfaces and increasing their hand-tomouth behavior. The level of exposure may continue to increase, given that the normal tendency ofyoung children to explore their environment orally increases through 2 years of age.

The actual dose to the child will depend on environmental concentrations and the efficiency of pesticide uptake for the different types of exposure routes such as dermal contact versus ingestion. To date, direct observation and quantification of children's exposure-related activity patterns and their interaction with their environment is very limited.

PROTECTING CHILDREN FROM PESTICIDES

Pesticides application is very important and useful to drive away pests in different homes and gardens but storing and application of this pesticide should be carefully applied away from children hence this could harm them. According to U.S. Environmental Protection Agency (USEPA) surveyed American households they found out that:

- Almost half of all homes with children under the age of 5 had at least one pesticide stored in an unlocked cabinet within a child's reach — that is, within 4 feet of the ground;
- About 75 percent of households without children under age 5 also stored pesticides in an unlocked cabinet within a child's reach and
- More than one in 10 pesticide poisonings happens in homes other than the child's home.

Storing pesticides properly: tips for protecting children

- Always store pesticides where children can't reach them or even see them;
- Put away pesticidesimmediately after use. If you are using a pesticide or other household chemical and are interrupted

(perhaps by a phone call), stop to properly close the container and remove it from a child's reach;

- Before applying pesticides (indoors or out), always take children and their toys, as well as pets, out of the area. Keep them away until the pesticide has dried or as long as the label advises;
- When applying insect repellents to children, read all directions first; do not apply over cuts, wounds or irritated skin; do not apply to eyes, mouth, hands or directly on the face. Use just enough to cover exposed skin or clothing, but do not use under clothing;
- Don't stockpile pesticides— buy only enough pesticide to carry you through the intended season. Storing pesticides 5 feet above ground level is important in homes with children under the age of 5;
- Garden pesticides should be locked in a cabinet or shed;
- Put other chemicals, such as kitchen and bathroom cleaners and chlorine bleach, behind locked doors as well;
- Purchase child-proof safety latches at a hardware store to install on cabinets;
- NEVER transfer pesticides to other containers that children may associate with food or drink. Every year, children are poisoned by pesticides careless adults placed in containers such as empty soda bottles and
- Make sure grandparents, babysitters and other people your child may visit know how to make their homes safe.

Creating a safer home

- Teach childrenhow to recognize and stay away from all poisonous materials, whether garden pesticides, household cleaners or prescription drugs;
- Always usehousehold products in child-resistant packaging. And use the packaging properly by closing the container tightly after you are done;
- Put away pesticides immediately after use. If you are using

a pesticide or other household chemical and are interrupted (perhaps by a phone call), stop to properly close the container and remove it from a child's reach.;

- Before applying pesticides (indoors or out), always take children and their toys, as well as pets, out of the area. Keep them away until the pesticide has dried or as long as the label advises;
- When applying insect repellents to children, read all directions first; do not apply over cuts, wounds or irritated skin; do not apply to eyes, mouth, hands or directly on the face. Use just enough to cover exposed;
- skin or clothing, but do not use under clothing and
- Don't stockpile pcDon't stockpile pesticides — buy only enough pesticide to carry you through the intended duration.

TREATMENT FOR PESTICIDE POISONING

Pesticides associated with each chemical class of pesticides such as. Organophosphates and carbamates are generally produce similar patterns of symptoms. The poisoning may be so mild that it can scarcely be detected, or it may become increasingly severe, depending upon the dose absorbed. One or more symptoms (headache, dizziness, nausea, etc.) may be common to many kinds of illnesses, whether caused by poisons or by viruses, bacteria, etc. It is not one or two symptoms, but the pattern of symptoms that makes it possible to tell one kind of poisoning from another.

Some manifestations to pesticide poisoning may lead to different sensations that only the applicator is aware of, such as nausea or headache; others, like an ashen skin color or rash, can be noticed by others.

There are two kinds of pesticide poisoning. Acute poisoning occurs after exposure to a single dose of pesticide. The appearance of symptoms may be sudden and dramatic or may be delayed, but usually occur within 12 hours. Chronic poisoning occurs as the result of repeated, small, nonlethal doses through a longer period of time. Many symptoms may appear such as nervousness, slowed reflexes, irritability or a general decline of health.

Figure 4.3: Some symptoms of pesticide poisoning in the body.

The following are general symptoms of pesticide poisoning:

Mild poisoning or early symptoms of acute poisoning:

Headache, fatigue, weakness, dizziness, restlessness, nervousness, perspiration, nausea, diarrhea, loss of appetite, loss of weight, thirst, moodiness, soreness in joints, skin irritation and irritation of eyes, nose and throat as shown in figure 5.3 common symptoms of mild poisoning.

Moderate poisoning or early symptoms of acute poisoning:

Nausea, diarrhea, excessive saliva, stomach cramps, excessive perspiration, trembling, no muscle coordination, muscle twitches, extreme weakness, mental confusion, blurred vision, difficulty in breathing, coughing, rapid pulse, flushed or yellow skin and uncontrollable crying as shown in figure 5.3.

Severe or acute poisoning:

Fever, intense thirst, increased rate of breathing, vomiting, uncontrollable muscle twitches, pinpoint pupils, convulsions, inability to breathe and unconsciousness. Medical doctors may be uninformed of the symptoms and treatment of pesticide poisoning because of the few cases which they treat. Commercial applicators or other large volume users should tell their doctor which of the pesticides (particularly the more toxic ones) they use, so the doctor will know the symptoms and the treatment and have antidotes on hand.

Since the late 1990's, the active ingredients in insecticides has changed from the organophosphate and carbamate insecticides to pyrethroid insecticides. The pyrethroid insecticides can cause a burning or prickling sensation of the skin and ingestion of some newer pyrethroids can cause seizures in small children and animals.

FIRST AID FOR PESTICIDE POISONING

A first aid assistance is needed while waiting for a medical help. But of course, an ambulance and a doctor are needed for a medical help for the victim. In this situation, make certain the victim is breathing and is not further exposed before leaving to make a phone call.

Be sure to take note of the label of the pesticide container or bring it to the doctor for examination. If this is not possible, make sure what pesticide the victim was using is known. Take note of the active ingredients statement on the label and write down the ingredients listed, as well as any information regarding treatment of persons exposed to the pesticide. Include also the registration number found either at the front or at the very end of the pesticide's label. While waiting for the doctor or ambulance to arrive, give first aid as follows:

Dilute Poison Quickly

First aid for a person who has swallowed a poison is to dilute the poison as quickly as possible by providing small sips of water – less than four (4) ounces and medical care without delay.

Warnings

- Never try to give anything by mouth to an unconscious victim and
- In an emergency, use any source of fairly clean water such as irrigation canals, lakes, ponds, watering troughs, etc. Don't let the victim die while worrying about dirty water.

Shock

Sometimes the poisoned victim can be in the state of shock and if this goes on untreated, the victim may lead to a fatal injury.

Steps to be followed if a person is poisoned:

Poison on the Skin

- The faster the poison is washed off the patient, the less injury will result;
- Remove clothing;
- Drench skin with water (shower, hose, faucet, pond);
- Cleanse skin and hair thoroughly with liquid detergent and water. (Detergents and commercial cleansers are better than soap.) and
- Dry and wrap the victim in a blanket.

Chemical Burns of the Skin

- Remove contaminated clothing;
- Wash with large quantities of running water;
- Immediately cover loosely with a clean, soft cloth and
- Avoid use of ointments, greases, powders and other drugs in first aid treatment of burns.

Pesticides in the Eye

- It is most important to wash the eye out as quickly, but as gently, as possible;
- Hold eyelids open and wash eyes with a gentle stream of clean running water.Continue washing for 15 minutes or more and
- Do not use chemicals or drugs in wash water. They may increase the extent of the injury.

Inhaled Poisons (Dusts, Vapors, Gases)

- If the victim is in an enclosed space, do not go in after them without an air-supplied respirator;
- Carry patient (do not let them walk) to fresh air immediately. Open all doors and windows;
- Loosen all tight clothing;

- Apply artificial respiration if breathing has stopped or is irregular. If the heart has stopped, perform cardiopulmonary resuscitation;
- Keep patient as quiet as possible;
- If patient is convulsing, watch their breathing and protect them from falling and striking their head. Keep their chin up so their air passage will remain free for breathing and
- Prevent chilling (wrap patient in blankets, but don't overheat). Do not give alcohol in any form.

Swallowed Poisons

- Poison Control and medical personnel do not recommend inducing vomiting and
- Usually the label will advise you in the First Aid Statement or Statement of Practical Treatment whether a person who has swallowed pesticide should be made to vomit. The Oklahoma Poison Control Center does not recommend inducing vomiting.

First Aid for Shock

- Unless the victim is vomiting, keep the victim flat on their back with their legs raised 1 to 1 ½ feet above their head level;
- Keep the victim warm enough to prevent shivering. Do no overheat;
- If the victim is conscious and has not swallowed any poison, give small sips of water and call 911 and
- Keep the victim quiet and reassure them often.

Long-term Health Effects of Pesticides

Most pesticide poisoning comes from contact with pesticides over weeks, months, or years, not from using them only once. People may not get sick from pesticides until many years later. In adults, it can take 5, 10, 20, 30 years or more to get sick from regular exposure. How long it takes for illness to show up depends on many things. In children, it usually takes less time. Illness from pesticides can start in a baby before the baby is born, while the mother is pregnant and in contact with pesticides.

When a person is exposed to pesticides over a long period of time, it is hard to know if his health problems are caused by pesticides. Long-term exposure may cause long-term harm, such as cancer, and damage to the reproductive system, the liver, the brain, and other parts of the body. Many long-term effects of pesticides are hard to see because people in farming areas are exposed to many different chemicals and because farm workers may move from place to place.

When people get cancer and other diseases, doctors and scientists may say the illness is due to chance, or to problems other than pesticides or contamination. They may tell us we cannot blame pesticides or other toxic chemicals. And sometimes people who sell pesticides or promote pesticide use will lie about it because they do not want to be responsible for other people's health problems. They can say this because it is often impossible to prove without a doubt that an illness which takes a long time to develop was caused by a particular pesticide or other toxic chemical.

SIGNS OF LONG-TERM ILLNESS FROM PESTICIDES

Pesticides and other toxics can cause plenty of chronic illnesses. Some signs of chronic illness are: weight loss, constant weakness, constant or bloody cough, wounds that do not heal, no feeling in the hands or feet, poor balance, loss of vision, very fast or very slow heartbeat, sudden mood changes, confusion, memory loss, and trouble concentrating.

Some long-term health effects of pesticides

Damage to the lungs

People exposed to pesticides may get a cough that never goes away, or have a tight feeling in the chest. These can be signs of bronchitis, asthma, or other lung diseases. Damage done over time to the lungs may lead to lung cancer. If you have any signs of lung damage, do not smoke cigarettes! Smoking makes lung disease worse.

Cancer

People exposed to pesticides have a higher chance of getting cancer than other people. This does not mean you will get cancer if you work with pesticides. But it means that working with pesticides gives you a higher risk of getting the disease. Hundreds of pesticides and pesticide ingredients are known or believed to cause cancer. Many more have not yet been studied. The most common cancers caused by pesticides are blood cancer (leukemia), non-Hodgkins lymphoma, and brain cancer.

Damage to the liver

The liver helps clean the blood and get rid of poisons. Because pesticides are very strong poisons, the liver sometimes cannot get rid of them. Severe liver damage can happen after a serious poisoning or after working with pesticides for many months or years.

Toxic hepatitis

It is a liver disease people get from being exposed to pesticides. Toxic hepatitis can cause nausea, vomiting and fever, yellowing of the skin, and can destroy your liver.

Damage to the nervous system

Pesticides damage the brain and the nerves. Long-term exposure to pesticides can cause loss of memory, anxiety, mood changes, and trouble concentrating.

Damage to the immune system

Some pesticides weaken the immune system, which protects the body from disease. When the immune system is weak from poor nutrition, pesticides, or from illnesses like HIV, it is easier to get allergies and infections and it is harder to heal from ordinary illnesses.

REPRODUCTIVE HEALTH EFFECTS OF PESTICIDES

Pesticides have many of the same reproductive health effects as other toxic chemicals. They can harm people's ability to have babies, or for babies to grow up healthy. Chemicals can enter a woman's body and appear in her breast milk later. There are so many pesticides in use all over the world that even mothers who have never used pesticides have some toxic chemicals in their breast milk.

Eventhough pesticide can be dangerous especially for breast feeding mother, but if you think your breast milk may have pesticides in it, the benefits of breastfeeding are stronger than any possible harm from pesticides in breast milk. Breast milk is the best food to help a baby grow healthy and strong.

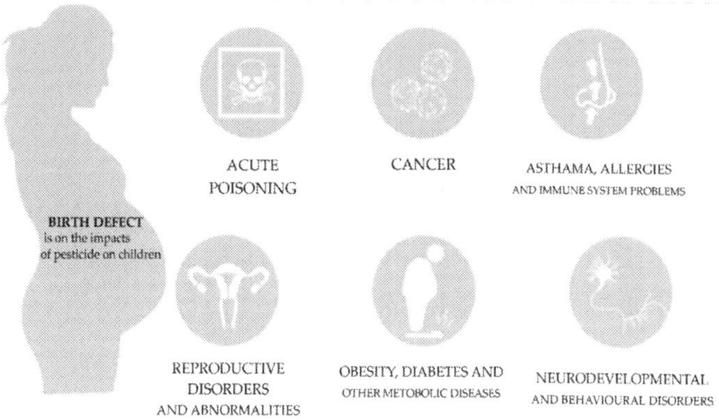

Figure 5.5: Some effects of pesticide in the reproductive health.

Some effects of pesticides on reproductive health are:

Sterility

Many male farm workers around the world have become unable to have children after they worked with certain pesticides because they can no longer make sperm.

Birth defects

When a pregnant woman is exposed to pesticides, the baby inside her is also exposed and can be damaged. Being exposed to pesticides when pregnant does not always mean the baby will have birth defects. But the baby will have a higher risk of having birth defects, learning difficulties, allergies, and other health problems as shown in figure 5.5.

Damage to hormone-producing glands

Hormones control many of our body activities, such as growth and reproduction. When pesticides damage the glands that produce hormones, this can cause problems with childbirth and reproduction. Even if a woman is exposed to pesticides before she is pregnant, she can have a miscarriage or the baby may be born dead because of the exposure.

REFERENCES

1. National Research Council, National Academy of Sciences. 1993. Pesticides in the Diets of Infants and Children, National Academy Press, Washington, DC. 184-185.
2. Bouchard et al. Pediatrics. 2010 Jun. www.ncbi.nlm.nih.gov/pubmed/204789452
3. Bradway D, Shafik T, Lores E. Comparison of cholinesterase activity, residue levels, and urinary metabolite excretion of rats exposed to organophosphorus pesticides. J Agric Food Chem 25:1353-1358 (1977).
4. CARB. Studies of Children's Activity Patterns. (Final Report) A733-149. Sacramento, CA: California Air Resources Board, 1991.
5. Eckerman DA, Gimenes LA, de Souza RC, Lopes Galvão PR, Sarcinelli PN, Chrisman JR. Age related effects of pesticide exposure on neurobehavioral performance of adolescent farm workers in Brazil. Neurotoxicol Teratol. 2007;29(1):164–175
6. Environmental advantages and disadvantages. Pediatrics. 2012 Nov. www.ncbi.nlm.nih.gov/pubmed
7. Environmental Working Group Shopper's Guide to Pesticides in Produce www.ewg.org/foodnews/summary/
8. FoodContaminantsAdulteration/PesticidesUS Dept of Agriculture http://www.nifa.usda.gov/Prog ViewOverview.cfm?prnum=1892
9. Forman J, Silverstein J; Comm on Nutrition; Council on Environmental Health. Organic foods: health and
10. Hill RH Jr, Head SL, Baker S, Gregg M, Shealy DB, Bailey SL, Williams CC, Sampson EJ, Needham LL. Pesticide residues in urine of adults living in the United States: reference range concentrations. Environ Res 71:99-108 1(1995). http://www.fda.gov/Food/FoodSafety/

11. Landrigan PJ, Claudio L, Markowitz SB, et al. Pesticides and inner-city children: exposures, risks, and prevention. Environ Health Perspect. 1999;107(suppl 3):431–437pmid:10346991
12. Lewis RG, Fortune CR, Blanchard FT, Camann DE. Movement and deposition of two organophosphorus pesticides within a residence after interior and exterior applications. J Air Waste Manag Assoc. 2001;51(3):339–351pmid:11266098
13. Loewenherz C, Fenske RA, Simcox NJ, Bellamy G, Kalman D. Biological monitoring of organophosphorus pesticide exposure among children of agricultural workers in central Washington State. Environ Health Perspect 105:1344-1353 (1997).
14. Lu C, Fenske RA, et al. Organic diets significantly lower children's dietary exposure to organo phosphorus pesticides. Environ Health Perspect. 2006 Feb. www.ncbi.nlm.nih.gov/pmc/articles / PMC1367841/
15. Lu C, Toepel K, Irish R, Fenske RA, Barr DB, Bravo R. Organic diets significantly lower children's dietary exposure to organophosphorus pesticides. Environ Health Perspect. 2006;114(2):260–263pmid:16451864
16. Michelson W. From Sun to Sun: Daily Obligations and Community Structure in the Lives of Employed Women and Their Families. Totowa, NJ:Rowman & Allanheld, 1985.
17. O'Rourke M. Personal Communication, 1998.
18. Shipp EM, Cooper SP, del Junco DJ, Bolin JN, Whitworth RE, Cooper CJ. Pesticide safety training among adolescent farmworkers from Starr County, Texas. J Agric Saf Health. 2007;13(3):311–321pmid:17892073
19. Silvers A, Florence BT, Rourke DL, Lorimor RJ. How children spend their time: a sample survey for use in exposure and risk assessments. Risk Anal 14:931-944(1994).
20. Steen W, Bond A, Mage D. Agricultural Health Study-Exposure Pilot Study Report. Research Triangle Park, NC:U.S. Environmental Protection Agency, 1997.
21. Stock T, Morandi M. A comparative evaluation of self-reported and independently-observed activity patterns in an air pollution health effects study. In: Unpublished presentation at Research Planning

Conference on Human Activity Patterns, Environmental Research Center, Las Vegas, Nevada, 1989.

22. US Environmental Protection Agency www.epa.gov/pesticides/US Food and Drug Administration
23. US EPA, Office of the Administrator, Environmental Health Threats to Children, EPA 175-F-96-001, September 1996. See also: http://www.epa.gov/pesticides/food/pest.htm.
24. USDA Choose My Plate www.choosemyplate.gov/food-groups/7 Find resources and information about pesticides on the Washington Toxics Coalition website at www.watoxics.org
25. WA State Dept of Health WIC Special Supplemental Nutrition Program for Women, Infants and Children
26. www.doh.wa.gov/YouandYourFamily/WIC.aspx

5
CHAPTER

ENVIRONMENTAL IMPACT OF PESTICIDES

INTRODUCTION

Pesticides are known to be one of the extremely useful and beneficial agents for preventing losses of crops as well as diseases in humans. Pesticides can have a multi-function which can be as destroying, repelling and mitigating agents. One of the negative feedback with the use of insecticides, insects , and pests may get an immunity with the constant use of commercial pesticides due to over usage especially pesticides have been developed which target multiple species.

Pesticides have proven to be a farmer's friend and tool to enhance production which was being applied by farmers around the world by increasing agricultural yield and by providing innumerable benefi ts to society indirectly. But this poses an issue on the ill effect of pesticides to human health and the environment has raised concerns about the safety of pesticides. Although we cannot diregard the problems associated with pesticide use, but we can circumvent them in one way or the other. Exposure to pesticides and hence the harmful consequences and ill- effects of this exposure can be minimized by several alternative way of cropping techniques or by using well-maintained spraying equipments. Production of better, safe and better pesticide formulations could mimimized the harmful effects associated with the pesticide usage. If the pesticides are used in appropriate quantities and used only when required or necessary, then pesticide risks can be minimized.

A less toxic formulation or lower dosage or low dose of a toxic formulation is used the effect can be solved. As Paracelsus also once said "The right dose differentiates a poison from a remedy". There are organochlorines, which are used as pesticides. Some pesticides are less biodegradable and in European countries had even banned the use of this pesticide with organochlorine compounds, it is only the developing countries are still using pesticides with organochlorines substances. This results in serious health hazards because this can be easily penetrate the water system and contributes to water pollution is on the rise due to these pesticides, even at low concentration, these pesticides have serious threat to the environment. The majority of farmers are unaware of the potential toxicities of pesticides. They have no information about types of pesticides, their level of poisoning, hazards and safety measures to be taken before use of those pesticides. Due to this reason, toxic and environmentally persistent chemicals are used to kill pests which can also lead to intentional, incidental or occupational exposure. These compounds have long term effects on human health.

Today, chemical application for pest eradication using either pesticide or insecticide as a chemical agent are becoming important in eliminating pests. To make the application of chemical pesticides a beneficial method for increasing yield in production, the integrated pest management protocol should be strictly followed to make it effective agent. On the level of population, the effects of pesticides will depend on the degree of exposure of the target and non-target organisms on the toxicity level, as well as on different factors like life history, characteristics, timing of application, population structure and landscape structure. The target insects actually are affected through its nerve that inhibits its action and metabolism.The neuroactive insecticides include acetylcholinesterase for organophosphates and methylcarbamates compounds, nicotinic acetylcholine receptors present in neonicotinoids, gamma-aminobutyric acid receptor channel for polychlorocyclohexanes and fiproles and voltage- gated sodium channels for pyrethroids and dichlorodiphenyltrichloroethane.

Pesticides used in the world are divided into different types depending on the target insect.And these are categorized as herbicides, insecticides, fungicides, rodenticides, molluscicides, nematicides that could enhance production as well as regulate the plant growth.But the application of pesticides should be regulated hence this could lead to environment disaster as a consequence.

Human health and biodiversity should be given the main concern on the overuse of pesticides, though economic side must be given a consideration. Pesticides can easily absorb in the environment hence it is more water soluble, heat stable and polar which makes it very difficult to reduce their lethal nature. Pesticides are not only toxic to people related to agriculture, but they can also affect in industries and public health work places. Depending upon the target species, pesticides can cause toxicities in natural flora, natural fauna and aquatic life.

PESTICIDES IN THE ENVIRONMENT

Transportation and Dispersion of Pesticides in Soil, Air and Water

Pesticides are being distributed to earth surface in the soil through different human activities such as agricultural ways by an application in farming, industry, and home uses in the environment. They are the contributing factor for the point source of pollution mainly discharges into the soil and transported to the different waterways by means of leaching. Thus, the discharge of the pesticides contaminated the earth surface then passes to the aquatic environment.

The main source of pollution count most in the agricultural sectors and most problematic sources of pollution will be washed out into the aquatic environments in any part of the world as seen in figure 5.1. Once in the environment, certain pesticides break down relatively quickly, whereas others persist over longer periods of time and can accumulate or transform into contaminant of different by-products. While the movement of pollution into the soil could be mobilized by means of precipitation and through wind process.

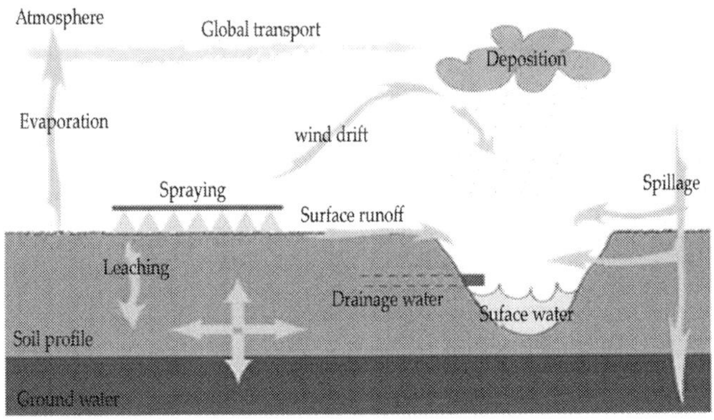

Figure 5.1: Pesticide spreading in the environment through different vectors.

Pesticide pollution is such a difficult problem to deal with due to the hazardous nature of the chemicals paired with the virtual inability to control their application. During application overspray chemicals are often carried in the wind to nearby ecosystems. Even in cases when agrochemicals are carefully applied, rinsing and subsequent runoff from rain, leeching into soil, and evaporation and subsequent deposition in the form of chemically tainted rain will cause traces of pesticides to spread far beyond their application sites and cause harm to non-target species.

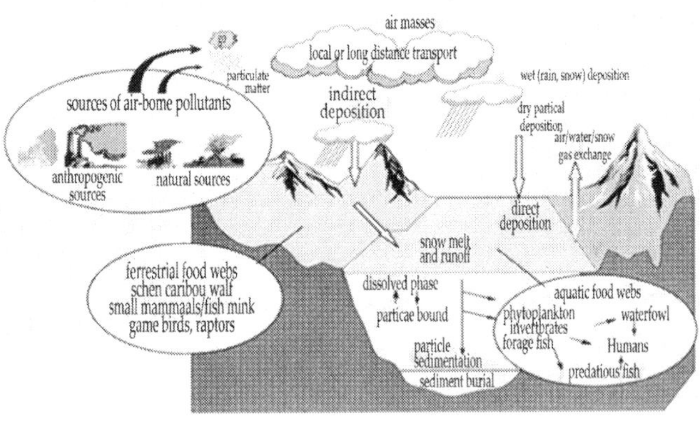

Figure 5.2: Long range transportation of pesticides especially the persistent Organic pollutants (POPs) in the environment

Pesticide residues can be carried out by the wind in the form of gas or vapor, particles or droplets that can be transferred in a long distance from their source. Contaminants can be deposited on the ground or in waterways where they may accumulate and/or transform. Figure 5.2 illustrates a typical path taken by certain pesticides in the environment. The contaminants are then concentrated on the plant's roots in the plant life and enter the food chain where they become concentrated in animal fat. The long-range cycle of transporting pesticide which has its chemical and physical characteristics which can be disastrous effect can be carried out in a long process the wind, waterways through leaching with its POP content can be pass on until the end of the Artic part. The POPs chemical found in the environment as a result of the application of pesticide through human activities can be classified in three forms such as semi-volatile, persistent and fat-soluble form. If it is in the form of semi-volatile form in a solid to the gaseous state at higher temperatures can be formed back into solid form at lower temperatures.

Persistent of the pesticides especially in southerly regions can evaporate and finally carried out by winds in effect condenses back due to cold air in the Arctic air that can be deposited on the ground. This pollutant can reach the Northern part of the pole and can accumulate hence the cooler temperature in the region prevents this from evaporating this gas. These POPs can go on as a cycle for several times at long distances in series of "hops" which led to the process being termed the "grasshopper effect" (Figure 5.3). POPs are persistent because they resist degradation under normal environmental conditions.

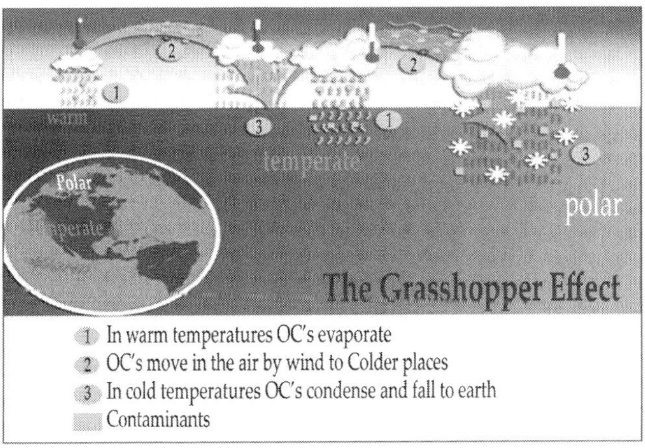

Figure 5.3: Long range transportation of pesticides especially the persistent Organic pollutants (POPs) in the environment

Fat soluble POPs, classified as one almost water insoluble form once ingested by living organisms cannot be excreted and tend to accumulate in the tissue of the organism. The bioaccumulation is a process wherein the poisonous compounds tend to remain in the living tissues of the organisms in a long period of time. As organisms continue to eat plants or animals which are being applied with pesticide tends to contaminate the organisms, and the concentration of contaminants increases as materials pass up the trophic levels of the food chain. This is known as biomagnification (Figure 5.4). POPs are known to biomagnify.

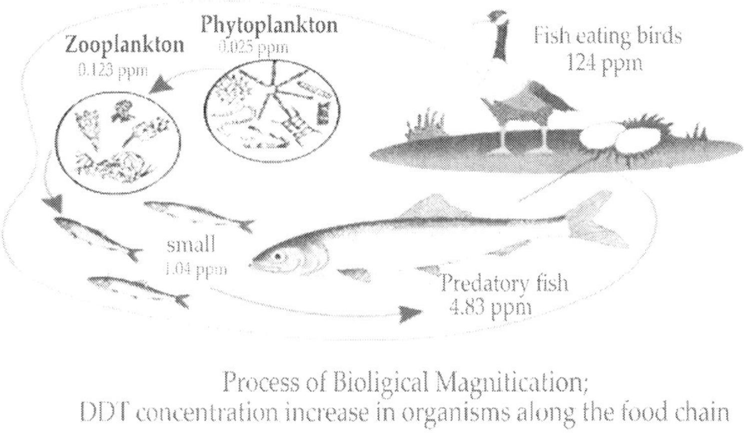

Figure 5.4: Long range transportation of pesticides especially the persistent Organic pollutants (POPs) in the environment

In the first place, organochlorine as one of the components in pesticides can be organic pollutants because of their persistence in the soil. This persistent is due to their characteristics of being semi-volatile compounds which are difficult ot degrade in normal environmental conditions. They cannot basically decide whether they want to be a solid, a fluid, or a gas.This can be explained in its chemistry principles because it is temperature dependent. So, at warm temperatures these substances will enter the atmosphere; at cold temperatures, they'll come down to the ground.

Pesticide in the form of Pos consist of fat soluble and having characteristics of a very low water-solubility. Though human bodies are make up to take out the toxins or pollutants converting them in the body systems into water-soluble parts that could be excreted as urine.

For fat-soluble substances, that mechanism doesn't work and they tend to build up over a lifetime. And could penetrate in the food chain due to the excessive application just to inhibit pests from the plants, which is the food source.

EFFECTS ON BIODIVERSITY

Biodiversity is thc process of the different way of life of the organism wherein it comprises the most complex form of organism with the interactions on how it interconnect and dependent with the environment. Biodiversity has the deeper significance in the functioning, nurturing and protecting different individual organisms which in turn has an interconnection with each other. The impacts that pesticides have on wildlife directly relates back into the functioning of biodiversity. The Earth's rich biological heritage of species, communities, and ecosystems, which have evolved over millions of years, is rapidly deteriorating and in many instances irreversibly disappearing. The aforementioned impacts that pesticides have on wildlife is a major cause for concern in regards to the deterioration of biodiversity.

Studies showed pesticides when applied to the aquatic environment may result in the loss of species diversity in the aquatic life and even predatory insects. In Europe, it has been found that a 42% loss in species abundance happens when it is exposed to pesticide, even when exposed to concentrations deemed environmentally safe by current legislation. Species richness of beneficial insects, such as bees, spiders, and beetles, has been found to be much higher on untreated or organic fields than fields treated with insecticides, which is a common occurrence in chemical-dependent agriculture.

Organic pest management sharply contrasts with a chemical-intensive approach in terms of its impact on the stability and resiliency of ecosystems. The effect of pesticide application magnifies its negative consequences for biodiversity and survival of wild species. Recognizing that various land management practices may have different effects on the web of life is crucial to maintaining the intricate balance and in sustaining life as a benefit in nature. Incorporating organic pest management over chemical-intensive controls is the most crucial change that can be made in order to mitigate negative impacts of pesticides to wildlife and to preserve the Earth's remaining biodiversity.

Pesticides get into the body of water through drifting, by runoff and most of all are through leaching in soil and even when directly applied through fumigation for mosquito control and inhibition of larvae hatching. Water contaminated with pesticide application can greatly affect any forms of life. It can affect aquatic plants, decrease dissolved oxygen in the water and can cause physiological and behavioral changes in fish populations. In several studies, lawn care pesticides have been found in surface waters and water bodies such as ponds, streams, and lakes.

Application of pesticides to the aquatic ecosystems through land drift can be toxic to fishes and non-target organisms. These pesticides are not only toxic but bring harm to fishes and other forms of aquatic life hence this is the contributory factor for algal blooms which losses biochemical oxygen demand among aquatic life. With the overuse of pesticides, this also increases algal bloom in effect results to decline in the different aquatic life especially fishes. Themselves but also interact with stressors which include harmful algal blooms.

The end result of pesticides on wildlife are extensive, and expose animals in urban, suburban and rural areas to unnecessary risks. Beyond pesticides considers wildlife to be any organism that is used for homes or either used in a lab. This includes but is not limited to bees, birds, small mammals, fish, other aquatic organisms, and the biota within soil. Wildlife can be impacted by pesticides through actual and indirect way, such as pesticide drift, secondary poisoning, runoff into local water bodies, and groundwater contamination. It is possible that some animals could be sprayed directly, while others consume plants or prey that have been exposed to pesticides.

Aquatic animals are exposed to pesticides in three ways (Helfrich et al. 2009).

- Dermally: Direct absorption via skin;
- Breathing: Uptake via gills during breathing and
- Orally: Entry via drinking contaminated water.

Figure 5.5: Pesticide residues coming from containers that can flow in the soil.

About 80 % of the dissolved oxygen is provided by the aquatic plants and it is necessary for the sustenance of aquatic life. The killing of aquatic plants by the herbicides results in drastically low O_2 levels and ultimately leads to suffocation of fish and reduced fish productivity).In all aspects of the levels of pesticides found in the land are much higher in surface waters comparing its presence in the groundwater probably because of surface runoff from farmland and contamination by spray drift. However, pesticides reach underground through seepage of surface water, improper disposal and accidental spills and leakages contaminated with the application of pesticides. Aquatic ecosystems are experiencing considerable damage due to drifting of pesticides into the body of waters like lakes, ponds, and rivers. Atrazine is known to be very hazardous to aquatic life especially fish species and it can directly and indirectly influence the immune systems of selected amphibians.

Amphibians are chiefly affected by pesticides contaminated surface waters, in addition to overexploitation and habitat loss .One type of pesticide carbaryl has been found hazardous in the different species of amphibians and another type of herbicideglyphosate is known to cause high mortality of tadpoles and juvenile frogs. Small concentrations of malathion have been shown to change the abundant presence of plankton and periphyton population has greatly affect the growth of frog tadpoles hence this could loss the oxygen air in the ponds and body of water. Moreover, chlorpyrifos and endosulfan also cause serious damage to amphibians.

Studies shows that about 10% of male frogs raised in atrazine-contaminated water developed into females. Male frogs that were genetically males phenotypically developed ovaries within their testes. They also developed the tendency to mate with other males and lay sustainable eggs and the reproductive potential of aquatic life forms also reduces due to herbicide spraying near weedy fish nurseries which eventually reduces the amount of shelter that is required by young fish to hide from predators (Helfrich et al, 2009).

Pesticide exposure can also moderately toxic to terrestrial plants in addition to killing non-target plants. Drifting or volatilization of phenoxy herbicides can injure nearby trees and shrubs. Herbicide glyphosate increases the susceptibility of plants to diseases and reduces seed quality. Even low doses of herbicides, sulfonylureas, sulphonamides and imidazolinones have a magnifying effect on the productivity of non-target crops, natural plant communities and wildlife.

Pesticides have not even spared the terrestrial animal populations wherein it is beneficial to insects such as bees and beetles can signifi cantly decline by the use of broad-spectrum insecticides such as carbamates, organophosphates and pyrethroids. Insect population has also been found to be greater on organic farms compared to non-organic ones. Synergistic effects of pyrethroids and triazole or imidazole fungicides are harmful to honey bees.

Neonicotinoids insecticides for example under this compounds are clothianidin and imidacloprid which are known to be lethal bees and swarms. Imidacloprid even at low dosage can have negative impact especially in the characteristics and behavioral capacities of bees and swarms. Wherein, one of behavioral sensitivities due to the neonicotinids are bees disappears in negatively affects bee foraging behaviour in addition to reducing learning capacity . The greatest havoc wreaked by neonicotinoids was the sudden disappearing of honey bees at the very start of the twenty first century. This was a major concern to the food industry as 1/3 of the food production depends on pollination by bees. Honey and wax obtained from commercial hives were reported to contain a mixture of pesticides of which neonicotinoids shared a significant portion. Since 2006, each year, honey bee populations have dropped by 29–36 % and pre-agricultural times, 20–25 % of the bird populations have declined.

One of the major causes of this massive decline is the use of pesticides which was not known before 1962. Pesticide accumulation in the tissues of bird species leads to their death. Bald eagle populations in the USA declined primarily because of exposure to DDT and its metabolites (Liroff, 2000). Fungicides can indirectly reduce birds and mammal populations by killing earthworms on which they feed. Granular forms of pesticides are disguised as food grains by birds. Organophosphate insecticides are highly toxic to birds and they are known to have poisoned raptors in the fields. Sublethal quantities of pesticides can affect the nervous system, causing behavioural changes (Pesticides reduce biodiversity, 2010).

For soil and crop plant, pesticide are applied by means of liquid spray and can be directly incorporated or injected into the soil or applied as granules or as a seed treatment. Once they have reached their target area, pesticides disappear via degradation, dispersion, volatilisation or leaching into surface water and groundwater; they may be taken up by plants or soil organisms or they may stay in the soil (Hayo & Werf, 1996) . The major concern of pesticide overuse is their leaching into the soil, which affects the microbes residing in it. Soil dwelling microbes help the plants in many different ways, such as nutrient uptake; breakdown of organic matter and increasing soil fertility. But indirectly they are also advantageous to humans as we heavily depend on plants. Unfortunately, pesticide overuse may have drastic consequences and a time may come when we would not have any more of these organisms and soil may degrade.

Several soil microbes are involved in the fixation of atmospheric nitrogen to nitrates. Chlorothalonil and dinitrophenyl fungicides have been shown to disrupt nitrifi cation and de-nitrifi cation bacteria dependent processes (Lang & Cai,2009). The herbicide, triclopyr inhibits soil bacteria involved in the transformation of ammonia into nitrite (Pell et al.1998) . Glyphosate, a non-selective herbicide, reduces the growth and activity of nitrogen-fi xing bacteria in soil (Santos & Flores, 1995) whereas, 2,4-D inhibits the transformation of ammonia into nitrates carried out by the soil bacteria (Frankenberger et al.1991).

Herbicides also cause considerable damage to fungal species in soil as pesticides trifl uralin and oryzalin both are known to inhibit the growth of symbiotic mycorrhizal fungi (Kelley and South 1978) that help in nutrient uptake. Oxadiazon has been known to reduce the number of fungal spores (Moorman,1989) whereas triclopyr is toxic to certain species of mycorrhizal fungi (Chakravarty & Sidhu,1987). Earthworms play a significant role in the soil ecosystem by acting as bio- indicators of soil contamination and as models for soil toxicity testing. Earthworms also contribute to soil fertility.

Pesticides have not spared earthworms from their toxic effects and the later is exposed to the former mainly via contaminated soil pore water. Schreck et al. (2008) reported that insecticides and/or fungicides produce neurotoxic effects in earthworms and after a long term exposure they are physiologically damaged (Schreck et al.2008).

Glyphosate and chlorpyrifos have deleterious effects on earthworms at the cellular level causing DNA damage. Glyphosates affect feeding activity and viability of earthworms (Casabé et al.2007). Goulson reviewed the harms of neonicotinoids on environment and animal life. He reported that as neonicotinoids have a tendency to accumulate in the soil, therefore, they can kill earthworms such as Eisenia foetida species (Goulson, 2013).

Pesticide exposure can be linked to cancer, endocrine disruption, reproductive effects, neurotoxicity, kidney and liver damage, birth defects and developmental changes in a wide range of species. Exposure to pesticides can also alter an organism's behavior, impacting its ability to survive. In birds, for example, exposure to certain pesticides can impede its singing ability, making it difficult to attract a mate and reproduce. Pesticides can also affect a bird's ability to care for its offspring, causing their young to die. For bees, even "near-infinitesimal" levels of systemic pesticides result in sublethal effects, impacting mobility, feeding behaviors, and navigation.

Many deformations have been found after exposure to hormone-mimicking pesticides classified as endocrine disruptors. The impacts of these chemicals include hermaphroditic deformities in frogs, pseudo-hermaphrodite polar bears with penis-like stumps, panthers with atrophied testicles, and intersex fish in rivers throughout the United States. Reproductive abnormalities have been observed in mammals, birds, reptiles, fish, and mollusks at levels considered "safe" by the Environmental Protection Agency (EPA).

Economic Impacts of Pesticides on Wildlife

Wildlife is used to insects, spiders, mammals, birds, fish, amphibians, reptiles, and plants. Each species fills a certain family and habitat which includes its specific food, cover, water, space, and breeding site preferences. The location where a species can meet all of its living requirements becomes that species' habitat. Wildlife habitats can be consideres as may be large and small, native and man-made, exist in urban settings, in agricultural fields, and in the wilderness.

Wildlife ecologists and natural resource managers are interested on the needs and the characteristics of wildlife especially its habits and habitats.The most specific concern of wildlife researchers is to know the life status of wildlife and to understand the different factors that affect there lives inorder to survive and be a viable populations. Most wildlife will adapt and flourish, given sufficient quantity of quality habitat, even in the presence of people. While ecological studies may pinpoint very specific requirements for individual species, the lives of plants and animals and their habitats can be integrated collectively into the ecosystem as a whole.

The biological and ecological inter-relationships of a plant from animals and how they co -exists as species plays in the ecosystem, is required to evaluate the potential impact of a specific pesticide on a *specific* species. The impact of a different pesticide it may be beneficial or has a negative effect to a species or its habitat as the chemical's residues move through the different parts of the system such as soil, water, food, or air. The interaction of wildlife, its habitat, and pesticides is evaluated by scientists trained in wildlife ecology, population dynamics, physiology, and environmental chemistry.

Pesticide Poisoning of Wildlife

Pesticides can be manifested in different forms depending on its use and the method of application such as applied into forests, rangeland, aquatic place, farmlots, rights-of-way, urban stations, and gardens. Their widespread use makes contact with pesticide residues inevitable for some wildlife. Pesticide poisonings to wildlife may result from acute or chronic exposure. Additionally, pesticides may affect wildlife either by secondary exposure or through indirect or direct effects to the animal or its habitat.

Pesticide application does not mean has only the negative ill effect to wildlife and also chemical residues does not necessarily lead to serious consequences for wildlife. The potential effect can be evaluated by simultaneously considering the availability of the pesticide and considering its degradation process, its lethal concentrations and most of all its ecological properties when one is expose to this type of pesticide. Due to the complexity of these issues, many scientific disciplines must play a role in both the studies and the interpretation of results. The results from scientific studies aid numerous federal and state natural resource agencies to assess and manage the effects of pesticides on wildlife, including endangered species.

Acute Poisoning

Exposure to pesticides can be dangerous to wildlife that may lead to death and cause of sickness to wildlife. Acute poisoning of wildlife may include fish that can be blamed to the application of pesticides wherein pesticides effluents can be pass through the different bodies of water such as ponds, streams, or rivers by surface runoff or spray drift, and bird die-offs caused by foraging on pesticide-treated vegetation or insects, or by consumption of pesticide-treated granules, baits, or seeds. These kind of poisoning can be identified and analyzed by getting the animal tissues through gas chromatography. The tissue of the fish or any aquatic animals can use to analyze the possible culprit of poisoninge or by investigating impacts on biochemical processes (e.g., cholinesterase levels in blood and brain tissue). In general, acute poisoning to wildlife takes place over a relatively short time, impacts a very localized geographical area, and is linked to a single pesticide.

Chronic Poisoning

Exposure of wildlife over an extended period of time to pesticide levels not immediately lethal may result in chronic poisoning. The most well-known example of a chronic effect in wildlife is that of the organochlorine insecticide DDT (via the metabolite DDE) on reproduction in certain birds of prey. DDT and other organochlorine pesticides such as dieldrin, endrin, and chlordane have been implicated in bird mortality resulting from chronic exposure. The reduction of these compounds in the 1970s, and early 1980s, has resulted in decreased organochlorine residues in most areas, and reproduction in birds, such as the bald eagle, has greatly improved. Organochlorine pesticides used in some foreign countries may pose risk to migratory birds which overwinter there.

Secondary Poisoning

Pesticides may impact wildlife through secondary poisoning when an animal consumes prey species that contain pesticide residues. Examples of secondary poisoning are (1) birds of prey becoming sick after feeding on an animal that is dead or dying from acute exposure to a pesticide, and (2) the accumulation and movement of persistent chemicals in wildlife food chains.

Indirect Effects

A pesticide may affect wildlife in ways other than direct or secondary poisoning. Pesticides may impact wildlife indirectly when a part of its habitat or food supply is modified. For instance, herbicides may reduce food, cover, and nesting sites needed by insect, bird, and mammal populations; insecticides may diminish insect populations fed on by bird or fish species; insect pollinators may be reduced, thereby affecting plant pollination. The study of indirect effects is an emerging area and one that may be difficult to investigate.

Organic Systems Protect Wildlife

One way to combat all of the negative impacts of pesticides to wildlife is to initiate organic practices on your own lawn and in your own garden while supporting organic agriculture instead of conventional agriculture that relies on pesticides. Beyond pesticides supports organic agriculture as effecting good land stewardship and reducing hazardous chemical exposures to wildlife. The pesticide pest management movement, indicating pesticide effects and problems associated with chemical agriculture, from groundwater contamination and runoff to drift, views organic as the solution to these serious environmental threats.

Conventional agriculture relies on a "pick and chooses" method when it comes to pesticide use, only treating the symptoms of bad land management instead of acknowledging the deeper problems and attempting to fully understand agriculture as a whole system, including impacts on wildlife. The greatest benefit and positive change would come from adopting a whole systems approach, starting with management methods which "feed-the-soil," promoting healthy land from the ground up. Beyond pesticides has long supported a "feed-the-soil" approach to agricultural management.Further, some programs are being instituted in promoting a system approach on the management of soil health and proper fertilization that eliminates synthetic fertilizers and focuses on building the soil food web and nurturing soil microorganisms. This approach will demonstrate on how to to enrich the soil with microbes that will produce a healthy land rich with nutrients that will benefit wildlife at the same time.

By promoting soil resiliency, the need for pesticides is dampened, which promotes natural predators and wildlife by giving them a chance to do what they were meant to do in nature. Organic systems could save wildlife from the effects of pesticides which encourages in the restoration of the natural balance and develop the conventional method in the agricultural systems.

EFFECTS OF PESTICIDES ON PLANTS AND ANIMALS

The most toxic absorption of pesticides can be absorbed through the skin swallowed or inhaled and this could be acquired when pesticides drift and settled on ponds laundry, toys, pools and furniture. People and pets track pesticides residue into the house and about 5% of pesticides reach target weeds especially when sprayed in the garden and farmlots. The rest runs off into water or condenses in the air while some drifts from landscaping ranges from 12 feet to 14.5 miles. More serious effect appear to be produced by direct inhalation of pesticide sprays than by absorption or ingestation of toxins.

Pesticides can be applied in different forms and methods, its application can be wide in terms of farm lots, aquatic environment, in forestland area, in urban communities and even directly applied to home gardens. The widespread use of pesticide has magnified the pesticide residues that inevitably affects wildlife that may result to acute or chronic exposure to animals or its habitat.

PESTICIDE IMPACT ON HUMAN HEALTH

The application of pesticide could not be denied that has improved the standard of human health by controlling vector-borne diseases, however, their long - term and indiscriminate use have resulted from side ill health effects. Many people especially infants and children are highly susceptible to deleterious effects of pesticides due to the non-specific nature and inadequate application of pesticides. As the pesticideapplication had tremendously increased over the past few decades, the likelihood of exposure to these chemicals has also increased considerably. World Health Organization specified that each year about 3,000,000 cases of pesticide poisoning and 220, 000 deaths are reported in developing countries. About 2.2 million people, mainly belonging to developing countries are at increased risk of exposure to pesticides. Besides, some people are more susceptible to the toxic effects of pesticide than others, such as infants, young children, agricultural farm workers and pesticide applicators.

Pesticides can penetrate in the the human body through the mouth by ingestion, in the air by inhalation or penetration via skin. But in most cases people are affected via the intake of pesticide contaminated food. After intake of pesticide in different pathway it will ultimately reach the human tissues and affects the major organs or can be stored in different compartments in the body. Although human bodies have mechanisms for the excretion of toxins, however, in some cases, it retains them through absorption in the circulatory system. Toxic effects are produced when the concentration of pesticide in the body increases far more than its initial concentration in the environment. The effects of pesticides on human health are highly variable. They may appear in days and are immediate in nature or they may take months or years to manifest and hence are called chronic or long-term effects. Acute and chronic effects of pesticide exposure on human health are discussed below.

Effects on Humans

The most important issue on the application of pesticides is in its effect on the environment, especially to human beings. Issues on the pesticides residues that can harm humans, aquatic life, and even plants is an increasing concern of the public. There are even increasing worries on the chemical residues borne with the use of pesticides and often suspected of being cancer- causing substances known as carcinogens that possible can inhibit endocrine activities which may be acquired in drinking water and food. Though there are different regulatory actions it maybe international and national regulatory agencies, reports of pesticide residues in human foods, both imported and home-produced, are numerous.

It was recorded that over the last fifty years most of human diseases and cause of deaths had resulted from exposure to different pesticides effects and about 20,000 deaths reported annually. Some deaths are due to accidents and suicidal cause but most involve some form of accidental exposure to pesticides especially among handlers of pesticides with farmers and spray operators in developing countries. Most of these cases are due to careless way and mishandling of applying pesticides without the proper protective gears and equipment. Moreover, there have been major accidents involving pesticides that have led to the death or illness of many thousands. One instance occurred in Bhopal, India, where more than 5,000 deaths resulted from exposure to accidental emissions of methyl isocyanate from a pesticide factory.

Effects on the Aquatic Environment

The use of agricultural chemicals is very important inorder to to control pests and increase yields in order to produce adequate food for the global population wherein it was estimated at 6.8 billion in 2009 was the gross production yield and 2015 reported to have reached 7 billion. Most of the underdeveloped countries, where 1.02 billion people (15 %) are estimated to lacking of food and classified undernourished and 1.3 billion people (19 %) live on an insufficient diet has the indire need of an adequate food supply.

This is the reason why agricultural production could not go away from the use of pesticide wherein, the agricultural sector's annual application of over 140 billion kilograms of fertilizers and large amounts of pesticides generates chemical residues that diffuse pollution of freshwater systems. The accumulation of these hazardous chemicals in both aquatic and terrestrial ecosystems has become an important issue globally. Researcher is looking for a solution of integrating a new type of pesticide that has less ill effects and less chemical residues that may harm human health, metals and many industrial chemicals interfere with the health and normal functioning of the endocrine systems of a wide range of organisms, including humans. It is believed that effects of these chemicals on the normal functioning of the endocrine system are responsible for a number of developmental anomalies in a wide range of species, from invertebrates to higher mammals.

One of the major environmental effects has been the possibility of fish kill which is happening in different regions in the world and marine invertebrates due to the contamination of aquatic systems by pesticides. This has resulted from the agricultural contamination of different body of water sources through fallout, drainage, or runoff erosion, and from the discharge of industrial effluents containing pesticides into waterways.

History will tell that even at early part of the century, most of the fish in Europe's Rhine River was killed due to pesticides discharges and the Great Lakes become contaminated with chemical residues from pesticides. In this effect, many organisms that help in the production of fishes and other aquatic life becomes susceptible to pesticides. This has been documented that it has indirect effects of pesticides on the fish food supply may have an even greater effect on fish populations. Some pesticides, such as pyrethroid insecticides, are extremely poisonous to most aquatic organisms. Therefore, investigative evidences shows that pesticides cause major losses in global fish production.

Acute Effects of Pesticides

Pesticides have been a dangerous effect to human health hazards which could range from short term effects with symptoms varies from headaches and nausea to chronic impacts like cancer, reproductive harm, and endocrine disruption. Acute impacts can give and ill-effects such as nerve, skin, and eye irritation and damage, headaches, dizziness, nausea, fatigue, and systemic poisoning - can sometimes be dramatic, and even occasionally fatal.

Other expected symptoms with pesticide exposure include headache, stinging of the eyes and skin, irritation of the nose and throat, skin disorders, appearance of the rash and blisters on the skin, dizziness, diarrhea, abdominal pain, nausea and vomiting, blurred vision, loss of sight and very rarely death. Acute effects of pesticide exposure are not severe enough for someone to seek medical help but should not go beyond the tolerable limit of exposure.

Chronic Effects of Pesticides

Chronic effects of pesticides are often lethal and may not appear even for years. These are long term effects that cause damage to multiple body organs .Pesticide exposure for prolonged periods of time results in following consequences:

- Pesticide exposure can cause a range of neurological health effects such as loss of coordination and memory, reduced visual ability and reduced motor signalling ;
- Long-term pesticide exposure damages the immune system and can cause hypersensitivity, asthma and allergies;
- Pesticide residues have been found in the bloodstream of cancer patients compared to normal individuals. Pesticides have been associated with leukaemia, brain cancer, lymphoma, cancer of the breast, prostate, ovaries, and testes;
- The presence of pesticides in the body for a longer time also affects reproductive capabilities by altering the levels of male and female reproductive hormones. Consequently, it results in stillbirth, birth defects, spontaneous abortion and infertility and
- Long-term exposure to pesticide also damages liver, lungs, kidneys and may cause blood diseases.

Ingestion of organochlorines causes hypersensitivity to light, sound, and touch, dizziness, tremors, seizures, vomiting, nausea, confusion and nervousness. Exposure to organophosphates and carbamates causes, symptoms similar to those of increased neurotransmitter-acetylcholine. These pesticides interfere with the normal nerve signal transduction and exposure to them causes headaches, dizziness, confusion, nausea and vomiting, muscle and chest pain. Difficulty breathing, convulsions, coma and death may occur in severe cases. Pyrethroids can cause an allergic skin response, aggressiveness, and hyper-excitation, reproductive or developmental effects in addition to causing tremors and seizures (Lah, 2011). It is observed that there is a relationship between pesticides and Parkinson's disease and Alzheimer's disease.

Farmers should be educated with the pesticide awareness should be arranged for these farmers to reduce the uses of toxic pesticides. In future chemical pesticides can be used in combination with natural treatments and remedies which result in more sustainable elimination of pests and insects. This combination not only promises environmental sustainability, but also has diverse applications in controlling of urban pests and invasive species. Pesticides have also posed a serious threat on biological integrity of marine and aquatic ecosystems. It is the need of time to integrate the studies of different disciplines including toxicology, environmental chemistry, population biology, community ecology, and conservation biology and landscape ecology to understand direct and indirect effects of pesticides on the environment.

REFERENCES

1. Agrawal A, Pandey RS, Sharma B (2010) Water pollution with special reference to pesticide contamination in India. J Water Res Prot 2(5):432–448
2. Aktar W, Sengupta D, Chowdhury A (2009) Impact of pesticides use in agriculture: their benefi ts and hazards. Interdiscipl Toxicol 2:1–12
3. Anon (1993) The environmental effects of pesticide drift, Peterborough: English Nature. 9–17. Benefi ts of pesticides and crop protection chemicals. In: Crop life America.
4. Brammall RA, Higgins VJ (1988) The effect of glyphosate on resistance of tomato to Fusarium crown and root rot disease and on the formation of host structural defensive barriers. Can J Bot 66:1547–1555
5. Casabé N, Piola L, Fuchs J, Oneto ML, Pamparato L, Basack S, Giménez R, Massaro R, Papa JC, Kesten E (2007) Ecotoxicological assessment of the effects of glyphosate and chlorpyrifos in an

Argentine soya fi eld. J Soils Sedim 7:232–239

6. Casida JE, Durkin KA (2013) Neuroactive insecticides: targets, selectivity, resistance, and secondary effects. Annu Rev Entomol 58:99–117
7. Chakravarty P, Sidhu SS (1987) Effects of glyphosate, hexazinone and triclopyr on in vitro growth of five species of ectomycorrhizal fungi. Eur J Pathol 17:204–210
8. Culliney TW, Pimentel D, Pimentel MH (1992) Pesticides and natural toxicants in foods. Agric Ecosyst Environ 41:297–320
9. Cunningham M (n.d.) Use of pesticides: benefi ts and problems associated with pesticides. In: Education portal. Available from http://education-portal.com/academy/lesson/use-of- pesticides-benefi ts-and-problems-associated-with-pesticides.html.
10. Damalas CA, Eleftherohorinos IG (2011) Pesticide exposure, safety issues, and risk assessment indicators. Int J Environ Res Public Health 8:1402–1419
11. Decourtye A, Lacassie E, Pham-Delègue MH (2003) Learning performances of honeybees (Apis mellifera L.) are differentially affected by imidacloprid according to the season. Pest Manag Sci 59:269–278
12. Delaplane KS (2000) Pesticide usage in the United States: history, benefi ts, risks, and trends. Cooperative Extension Service. The University of Georgia, College of Agricultural and Environmental Sciences. Bulletin 1121. Reprinted November, 2000. http://pubs.caes.uga.edu/ caespubs/pubs/PDF/B1121.pdf
13. Dreistadt SH, Clark JK, Flint ML (1994) Pests of landscape trees and shrubs. An integrated pest management guide. University of California Division of Agriculture and Natural Resources. Publication No. 3359
14. Environmental impacts. In: Pesticide Action Network. Available from http://www.panna.org/ issues/persistent-poisons/environmental-impacts.
15. FAO (2002) Manual on the submission and evaluation of pesticide residues data for the estimation of maximum residue levels in food and feed. Food and Agriculture Organization, Rome

16. Fletcher JS, Pfl eeger TG, Ratsch HC (1993) Potential environmental risks associated with the new sulfonylurea herbicides. Environ Sci Technol 27:2250–2252
17. Forson DD, Storfer A (2006) Atrazine increases Ranavirus susceptibility in the tiger salamander (Ambystoma tigrinum). Ecol Appl 16:2325–2332
18. Frankenberger WT, Tabatabai MA Jr, Tabatabai MA (1991) Factors affecting L-asparaginase activity in soils. Biol Fert Soils 11:1–5
19. Gentz MC, Murdoch G, King GF (2010) Tandem use of selective insecticides and natural enemies for effective, reduced-risk pest management. Biol Control 52(3):208–215
20. Goulson DJ (2013) An overview of the environmental risks posed by neonicotinoid insecticides. J Appl Ecol 50:977. doi: 10.1111/1365-2664.12111
21. Hayo MG, Werf VD (1996) Assessing the impact of pesticides on the environment. Agric Ecosyst Environ 60:81–96
22. Helfrich LA, Weigmann DL, Hipkins P, Stinson ER (2009) Pesticides and aquatic animals: a guide to reducing impacts on aquatic systems. In: Virginia Polytechnic Institute and State University. Available from https://pubs.ext.vt.edu/420/420-013/420-013.html.
23. Hicks B (2013) Agricultural pesticides and human health. In: National Association of Geoscience Teachers. Available from ttp://serc.carleton.edu/NAGTWorkshops/health/case_studies
24. History of pesticide use (1998) Available from http://www2.mcdaniel.edu/Biology/eh01/pesticides /historyofpesticidesuse.html.
25. Public Health. Available from http://www.peelregion.ca/health/topics/pesticides/why-reduce/why-reduce4.htm.
26. Pesticides in Groundwater (2014) In: The USGS Water Science School. Available from http:// water.usgs.gov/edu/pesticidesgw.html.
27. Jabbar A, Mallick S (1994) Pesticides and environment situation in Pakistan (Working Paper Series No. 19). Available from Sustainable Development Policy Institute (SDPI) Jacobs A (n.d.). Timeline-history of pesticides. In: Pesticides & University of Oregon.

Available from http://blogs.uoregon.edu/ajacobssu13gateway/timeline/ +Health .

28. Lang M, Cai Z (2009) Effects of chlorothalonil and carbendazim on nitrifi cation and denitrifi cation in soils. J Environ Sci 21:458–467
29. Liroff RA (2000) Balancing risks of DDT and malaria in the global POPs treaty. Pestic Safety News 4:3 Locke D, Landivar JA, Moseley D (1995) The effects of rate and timing of glyphosate applications of defoliation effi ciency, regrowth inhibition, lint yield, fi ber quality and seed quality. Proc Beltwide Cotton Conf 2:1088–1090
30. Macneale KH, Kiffney PM, Scholz NL (2010) Pesticides, aquatic food webs, and the conservation of Pacifi c salmon. Front Ecol Environ 8:475–482 M ajewski M, Capel P (1995) Pesticides in the atmosphere: distribution, trends, and governing factors. Pesticides in the hydrologic system, vol 1. Ann Arbor Press Inc., Boca Raton, FL, p 118
31. Monaco JT, Weller SC, Ashton FM (2002) Herbicide registration and environmental impact. Weed science: principles and practices, 4th edn. John Wiley & Sons, New York, NY
32. Moorman TB (1989) A review of pesticide effects on microorganisms and microbial processes related to soil fertility. J Prod Agric 2:14–23
33. Pell M, Stenberg B, Torstensson L (1998) Potential denitrifi cation and nitrifi cation tests for evaluation of pesticide effects in soil. Ambio 27:24–28
34. Pesticides 101-A Primer. In: Pesticide Action Network North America. Available from http:// www.panna.org/issues/pesticides-101-primer.
35. Pesticides and Human Health. In: Californians for Pesticide Reforms. Available from http://www. pesticidereform.org/section.php?id=38.
36. Pesticides. In: GRACE Communications Foundation. Available from http://www.sustainabletable. org/263/pesticides.
37. Pilling ED, Jepson PC (2006) Synergism between EBI fungicides and a pyrethroid insecticide in the honeybee (Apis mellifera). Pestic Sci 39:293–297

38. Rashid B, Husnain T, Riazuddin S (2010) Herbicides and pesticides as potential pollutants: a global problem. Plant adaptation phytoremediation. Springer, Dordrecht, pp 427–447
39. Relyea RA (2005) The lethal impact of roundup on aquatic and terrestrial amphibians. Ecol Appl 15:1118–1124
40. Relyea RA, Hoverman JT (2008) Interactive effects of predators and a pesticide on aquatic communities. Oikos 117:1647–1658
41. Rohr JR, Schotthoefer AM, Raffel TR, Carrick HJ, Halstead N, Hoverman JT, Johnson CM, Johnson LB, Lieske C, Piwoni MD, Schoff PK, Beasley VR (2008) Agrochemicals increase trematode infections in a declining amphibian species. Nature 455:1235–1239
42. Ross G (2005) Risks and benefi ts of DDT. Lancet 366(9499):1771–1772 Santos A, Florcs M (1995) Effccts of glyphosatc on nitrogcn fi xation of free-living heterotrophic bacteria. Lett Appl Microbiol 20:349–352
43. Schmolke A, Thorbek P, Chapman P, Grimm V (2010) Ecological models and pesticide risk assessment: current modeling practice. Environ Toxicol Chem 29(4):1006–1012
44. Scholz NL, Fleishman E, Brown L, Werner I, Johnson ML, Brooks ML, Mitchelmore CL (2012) A perspective on modern pesticides, pelagic fi sh declines, and unknown ecological resilience in highly managed ecosystems. Bioscience 62(4):428–434
45. Schreck E, Geret F, Gontier L, Treilhou M (2008) Neurotoxic effect and metabolic responses induced by a mixture of six pesticides on the earthworm Aporrectodea caliginosa nocturna. Chemosphere 71(10):1832–1839
46. Sharma DR, Thapa RB, Manandhar HK, Shrestha SM, Pradhan SB (2012) Use of pesticides in Nepal and impacts on human health and environment. J Agric Environ 13:67–72
47. Sparling DW, Feller GM (2009) Toxicity of two insecticides to California, USA, anurans and its relevance to declining amphibian populations. Environ Toxicol Chem 28(8):1696–1703 Spear R (1991) Recognised and possible exposure to pesticides. In: Hayes WJ, Laws ER (eds) Handbook of pesticide toxicology. Academic, San Diego, CA, pp 245–274

48. Speck-Planche A, Kleandrova VV, Scotti MT (2012) Fragment-based approach for the in silico discovery of multi-target insecticides. Chemom Intell Lab Syst 111:39–45 Straathoff H (1986) Investigations on the phytotoxic relevance of volatilization of herbicides. Mededelingen 51(2A):433–438
49. The Asian Amphibian Crisis (2009) In: IUCN. Available from http://www.iucn.org/about/union/ secretariat/offi ces/asia/regional_activities/asian_amphibian_crisis/
50. Van Djik TC (2010) Effects of neonicotinoid pesticide pollution of Dutch surface water on non- target species abundance
51. Warsi F. How do pesticides affect ecosystems. In: Pesticides. Available from http://farhanwarsi. tripod.com/id9.html.
52. WHO (2010) International code of conduct on the distribution and use of pesticides: guidelines for the registration of pesticides. World Health Organization, Rome
53. Yang EC, Chuang YC, Chen YL, Chang LH (2008) Abnormal foraging behavior induced by sublethal dosage of imidacloprid in the honey bee (Hymenoptera: Apidae). J Econ Entomol 101:1743–1748
54. Bordás I., 2006, Veszélyes anyagok, készítmények, peszticidek. Országos Kémiai Biztonsági Intézet
55. Dési I., Márton M., Gönczi Cs.-né, Páldy A., Király O.-né, Varga Gy.-né, 1983, A lakosság és a táj vizsgálatának jelentősége a peszticidek okozta megbetegedések és a környezetkárosítás megelőzésében. Földrajzi közlemények 3-4, 309-315.
56. MethodEx Policy Toolbox, 2007, BeTa-MethodEx. In MethodEx, 2007, Methods and data on environmental and health externalities: harmonising and sharing of operational estimates. Final Technical Report: Methods. pp.299.
57. Pearce, D., W., Markandya, A. and Barbier, E., B., 1989, Blueprint for a Green Economy. Earthscan Publication Ltd, London. pp.192
58. Pretty, J.N.; Brett, C.; Gee, D.; Hine, R.E.; Mason, C.F.; Morison, J.I.L.; Rayment, M.D.; van der Bijl, G.; Dobbs, T, 2001, Policy and Practice. Policy Challenges and Priorities for Internalizing the Externalities of Modern Agriculture. Journal of Environmental Planning and Management, 44(2), 263–283.

59. Travisi, C. M. and Nijkamp, P., 2008, Valuing environmental and health risk in agriculture: A choice experiment approach to pesticides in Italy. Ecological Economics, Volume 67, Issue 4, 1 November 2008, p.598-607.

6 CHAPTER

APPLICATION OF CHEMICALS IN THE ENVIRONMENT

INTRODUCTION

Since the publication of Rachel Carson's landmark 1962 book Silent Spring, the impacts of pesticides on the environment have been well known. Pesticides are toxic to living organisms. Some can accumulate in water systems, pollute the air, and in some cases have other dramatic environmental effects. Scientists are discovering new threats to the environment that are equally disturbing. Pesticide use can damage agricultural land by harming beneficial insect species, soil microorganisms, and worms which naturally limit pest populations and maintain soil health. Further, this could affect the growth of plants by weakening its immune systems especially the plant roots which is the basic tool for plant food production and this also reduces concentrations of the important nutrients needed by the plant taken into the soil for nitrogen and phosphorous content.

PESTICIDE USE AND THE ENVIRONMENT

Pesticide applications cannot be safely undertaken without an understanding of the interconnections among air, water, land, and the life they support. By understanding these environmental connections, pesticide applicators can better manage target pests while minimizing harm to the environment.

Figure 6.1: Pesticides on air contributes air pollution

Air

Pesticides can also be found in rain, ground water, streams, rivers, lakes and oceans. There are 4 major ways that pesticides can reach the water: it can drift outside of the area of where was sprayed, it may leach through the soil, it could be carried as runoff, or it may be spilled accidentally. Some studies showed that pesticide concentrations exceed those allowable for drinking water in some samples of river water and groundwater. When pesticides are applied, they may be carried off-site in the air as spray droplets, vapors, solid particles, or even treated soil particles. The physical movement of pesticide spray droplets through the air, at the time of or soon after the pesticide application, to any site other than the intended site is known as pesticide drift. Pesticides applied in windy or dead calm conditions (during a temperature inversion) can move off-site and harm humans, animals, and the environment.

Water

Figure 6.2: Pesticides in water can cause water pollution and causes death to aquatic species.

Actually pesticides are monitored in lakes, rivers, streams, groundwater, and rain by some government agencies especially on the water quality program. The majority of these detections are at very low levels especially in areas wherein there is a ramphant application of pesticide near farmlots and agricultural areas. However, a few pesticides have been detected at levels approaching or exceeding water quality benchmarks set for protection of humans or aquatic organisms. As pesticides move through the ground, microorganisms in the soil can break down the pesticide into other chemicals. Pesticides can also be broken down by sunlight, naturally occurring chemicals in the soil, and by plants. Most of these break-down products are less toxic to the environment than the original pesticide, but some can be more toxic.

Land

Land consists of different types of soil such as sand, clay, or silt which are often layered on top of one another, underlying or exposed rock formations, and groundwater. Pesticides properly applied to the land surface as directed by the label usually stay there until they break down into inactive compounds. Problems arise when surface runoff carries pesticides into lakes, streams, and wetlands; or water percolating through the soil carries pesticides down to groundwater. Pesticides with a long effective half-life break down slowly and may move to a location where they could harm plants, insects, animals, or bacteria. The use of pesticides decreases the general biodiversity in the soil. If there is no chemicals in the soil there is a higher soil quality, and this allows for higher water retention, necessary for plants to grow.

Figure 6.3: Pesticides in soil and land that can leach out to the body of water.

PESTICIDE FATE IN THE ENVIRONMENT

Several processes affect the fate of pesticides in the environment. These include the way pesticides move in air, water, or soil after they are applied through transportation and how they change after interacting with the soil, plants, or sunlight by transformation.Among the ways that pesticides can be transported are:

- Spray drift –during application, pesticides in spray droplets can be shifted by the wind from the location where they were applied to another location;
- Volatilization –pesticides can evaporate into the air and then the vapors may move on the wind.

- Runoff – pesticides can run off the land during rainfall, snowmelt, or irrigation;
- Leaching – pesticides can percolate through the soil into groundwater
- Adsorption – pesticides can attach to soil and then move when soil is eroded by water or wind and
- Transformation often involves the process of decomposition, when the original pesticide chemistry is transformed into a different, usually less complex, chemistry. As the break-down process progresses, the original pesticide disappears from the environment and is replaced by a series of break-down products, also called degradates.

Pesticides can be transformed by:

- Plant uptake –plants can take up pesticides from the soil.
- Photodecomposition –sunlight can break down pesticides.
- Microbial decomposition –microorganisms in the soil can break down pesticides into different chemical compounds.
- Chemical degradation –over time, the soil's chemistry can break down or change as bonds between molecules creating new compounds.

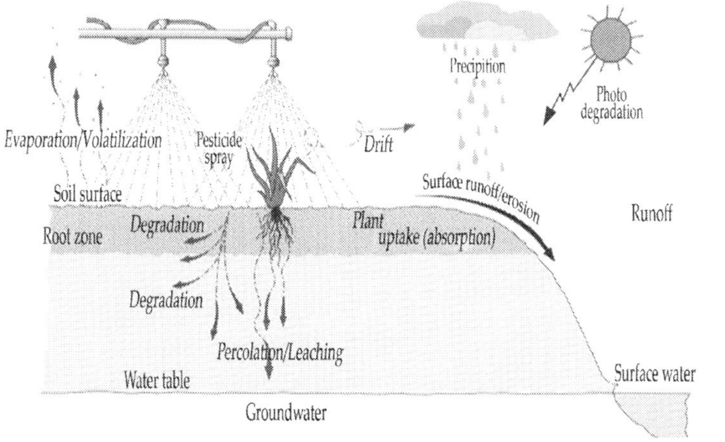

Figure 6.4: Fate of pesticides in the environment. (Adapted from the University of Illinois General Standards manual)

PESTICIDE PROPERTIES THAT AFFECT ITS ENVIRONMENTAL FATE

Pesticides' chemical properties affect not only how effective they are, but also how they may move, persist, or break down in the environment In other words, a pesticide's specific properties affect its fate in the environment.

Pesticides that are volatile which can easily evaporate readily at normal temperatures and pressures are more likely to end up drifting to non-target areas. Pesticides that have high solubility which means that the amount of a substance that will dissolve in a given amount of another substance, such as water, at a specified temperature and pressure are more likely to end up in surface water. Highly soluble pesticides can mix as easily with rain as with the water used to dilute them to the appropriate application strength. Pesticides with high solubility also percolate into groundwater more rapidly than less soluble compounds.

Pesticides with high adsorption can adhere in an extremely thin layer of molecules to the surfaces of soil or sediment particles and stick with the soil and stay in the environment longer. Pesticides with a long soil half-life that is the time required for half the amount of a substance introduced into a living system or cosystem to be eliminated or disintegrated by natural processes which stay in the soil long after one or many treatments. Over time the pesticide concentration can actually build up in the soil.

ENVIRONMENTAL FACTORS THAT AFFECT PESTICIDE FATE

Weather conditions have a major impact on pesticide fate. Rain, snowmelt, wind, humidity, temperature, sunlight, and temperature inversion all influence how pesticides are transported or transformed.

- ***Temperature inversions*** occur when the layer of air at ground level is colder than air higher up—the opposite of normal air temperatures. Inversions are very stable and create "dead calm."

Spray droplets from pesticides applied during inversions can stay in a concentrated cloud and move slowly toward sensitive plants and animals.

- ***Temperature increases*** make pesticides more volatile and may cause pesticides to move as a vapor to non-target sites and harm sensitive plants and animals. That is why some pesticide labels restrict application over a certain temperature. Soil properties can influence how quickly chemical pesticides break down or move from the application location.
- ***Organic matter and clay particles*** in the soil are good at tying up (adsorbing) pesticides to their surfaces so the pesticide is no longer available in the soil. This is why some pesticides labels have higher application rates on soils high in organic matter or clay: to ensure effectiveness. When adsorbed to the surface of organic matter and clay particles, pesticides are prevented from moving to contaminate groundwater or surface water. Organic matter in soils also contains nutrients that nourish micro-organisms that naturally break down pesticides.
- ***Soil texture and porosity*** also influence pesticide fate. Soils that are "tight," such as those with lots of clay, have the benefit of preventing leaching, and so pesticides are not likely to reach groundwater. But such soils also allow for greater pesticide volatilization and surface runoff. Water moves rapidly through soils that are "coarse textured," such as sandy or gravelly soil, increasing the chance of pesticides leaching to groundwater, but preventing surface runoff.

Finally, a soil that is steeply sloped allows dissolved pesticides to run off more rapidly than soils on flat or gently sloping lands. The diagram below shows the breakdown of pesticides in the environment.

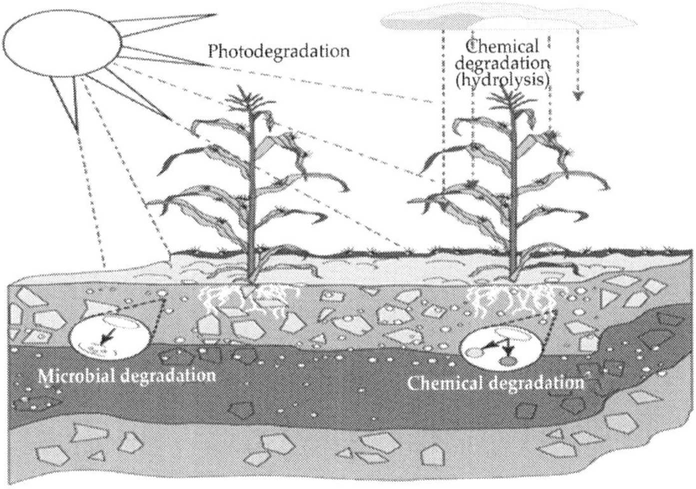

Figure 6.5: Breakdown of pesticides in the environment (National Pesticide Applicator Certification Core Manual. Adapted from OSU Pesticides and Groundwater Contamination)

PESTICIDE IMPACTS ON ECOSYSTEMS

An ecosystemis formed by the interaction of a community of organisms with their environment. When a pesticide is introduced, it can disrupt an ecosystem in a number of ways:

- It repels, controls, or kills the target pest The target pest, however, may be food for another organism. If the target pest is the major food source for a local bird species, the unintended outcome of killing the pest may be reducing the bird population.
- A pesticide that effectively kills a target organism may also effectively kill a beneficial species. Insecticides that kill wasps in a barn, for example, may also kill a hive of bees needed to pollinate fruit trees.
- Pesticides can accumulate through the food chain. As multiple small organisms are eaten by larger ones, pesticides found in the smaller organisms can build up in the tissues of the larger one. Biomagnificationis a process whereby some organisms

accumulate chemical residues in higher concentrations than those found in the organisms they consume as shown in figure 7.6.

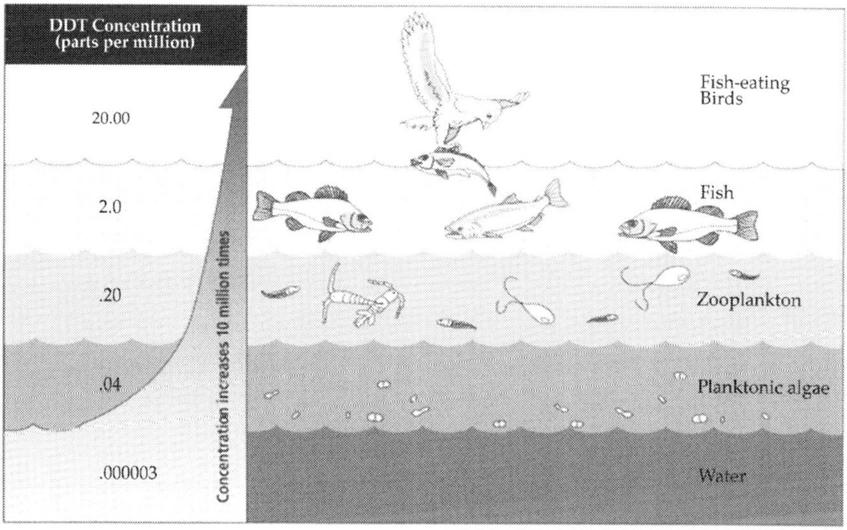

Figure 6.6: Biomagnification in the food chain. (Adapted from Penn. State Pesticide Education Manual).

Sensitive Sites

Pesticide applicators exercise caution at any treatment location, but certain sensitive sitesrequire greater planning and care. As the name indicates, sensitive sites are more susceptible to negative side effects that pesticides may have. Sensitive sites include:

Areas producing organic crops or livestock. Products certified as "organic" are required to be grown free from most man-made pesticides for three or more years. Applying a pesticide not approved for organic production that ends up in the soil, water, or groundwater of an organic operation could have major economic and legal consequences.

- **Lawns and gardens**. Drift minimization is especially important here. Children and pets play on lawns. People of all ages and states of health eat food grown in gardens.
- **Lakes, rivers, and wetlands.** Many pesticides are toxic to fish and other aquatic organisms, including aquatic plants that

form the base of the food chain. Minimize drift, leaching, and runoff.

- **Parks and protected natural areas**. Parks and natural areas contain high-value environmental resources, including threatened or endangered species. It is illegal to harm an endangered or threatened species, or to destroy or modify its critical habitat, when using a pesticide. Protect people in these areas, too, by observing safe application methods.
- **Drinking water supplies.** We all need safe drinking water. Applicators should know if a treatment site has any wells or is within a wellhead protection area (area around a well from which drinking water is drawn). Most at risk are shallow wells, abandoned or broken wells that may be hidden, and wells located in very permeable soil.
- **Municipal drinking water well fields**. Location and depth of municipal drinking water wells is an important consideration. Check with the local municipal water supplier if there is any possibility that the treatment site may be within a wellhead protection area. This is not always easy to determine without consulting the municipal water supplier.
- **Waste management areas**. Bacteria play an essential role in breaking down wastes in compost and manure treatment facilities, and pesticides can interfere with or kill beneficial bacteria. In addition, finished compost and re-applied manure are both recycled into the food supply.
- **Air intake systems**. Airborne pesticides that drift into an air intake system can be rapidly distributed throughout a building.
- **Daycares, schools, senior facilities, or health-care facilities.** These sensitive sites are likely to house children, the elderly, or people with health issues, all of whom may be more sensitive to pesticides than healthy adults.

REFERENCES

1. Shiklomanov, I.A. (1998). World Water Resources: A new Appraisal and Assessment for the 21st Century. An IHP report, UNESCO,

Paris, France.

2. Moiseenko, T.I. (2008). Aquatic ecotoxicology: theoretical principles and practical application. Water Resources, 35(5), 530-541.
3. Vorosmarty, C.J., McIntyre, P.B., Gessner, M.O., Dudgeon, D., Prusevich, A., Green, P., Glidden, S., Bunn, S.E., Sullivan, C.A., Liermann, C.R. and Davies, P.M. (2010). Global threats to human water security and river biodiversity. Nature, 467, 555-561.
4. Schwarenbach, R.P., Escher, B.I., Fenner, K., Hoffstetter, T.B., Johnson, C.A., von Gunten, U. and Wehrli, B. (2006). The challenge of micropollutants in aquatic systems. Science, 313, 1072-1077.
5. Jurado, A.S., Fernandes, M.A.S., Videira, R.A., Peixoto, F.P. and Vicente, J.A.F. (2011). Herbicides: The Face and the Reverse of the Coin. An in vitro Approach to the Toxicity of Herbicides in Non-Target Organisms, Herbicides and Environment, Andreas Kortekamp (Ed.), ISBN: 978-953-307-476-4, InTech. http://www.intechopen.com/articles/show/title/herbicides-the-face-and-the-reverse-of-the-coin-an-in-vitro-approach-to-the-toxicity-of-herbicides-i. (Accessed September, 2013).
6. PRB (Population Reference Bureau) (2012). 2012 World population data sheet, published by the Population Reference Bureau, 1875 Connecticut Ave., NW, Suite 520, Washington, DC 20009 USA, http://www.prb.org/pdf12/2012-population-data-sheet_eng.pdf, (accessed December 2012).
7. LeBlanc, G.A. (2007). Crustacean endocrine toxicology: a review. Ecotoxicology, 16, 61-81.
8. Correia, T.G., Narcizo, A.M., Bianchini, A., and Moreira, R.G. (2010). Aluminum as an endocrine disruptor in female Nile tilapia (Oreochromis niloticus). Comparative Biochemistry and Physiology, Part C, 151, 461-466.
9. Pedersen, M., Halldorsson, T.I., Mathiesen, L., Mose, T., Brouwer, A., Hedegaard, M., Steffen, L., Kleinjans, J.C.S., Besselink, H. and Knudsen, L.E. (2010). Dioxin-like exposures and effects on estrogenic and androgenic exposures and micronuclei frequency in mother–newborn pairs. Environment International, 36, 344-351.

10. Abel P.D. (2002). Water Pollution Biology (2nd edn.), Taylor and Francis Ltd, London.
11. London, L., Dalvie, M.A. and Cairncross, E. (2005). Approaches for regulating water in South Africa for the presence of pesticides. Water SA, 31(1), 53-59
12. Mihaich, E.M., Friederich, U., Caspers, N., Hall, A.T., Klecka, G.M., Dimond, S.S., Staples, C.A., Ortego, L.S. and Hentges, S.G. (2009). Acute and chronic toxicity testing of bisphenol A with aquatic invertebrates and plants. Ecotoxicology and Environmental Safety, 72, 1392-1399.
13. Benstead, R.S., Baynes, A., Casey, D., Routledge, E.J., Jobling, S. (2011). 17 β-Oestradiol may prolong reproduction in seasonally breeding freshwater gastropod molluscs. Aquatic Toxicology, 101, 326-334.
14. Lipika, P. and Patra, A. K. (2006). Haemoatopoietic alterations induced by carbaryl in Clarias batrachus (LINN), Journal of applied Sciences and Environmental Management, 10 (3), 5-7.
15. Boran, M., Altinok, I., Capkin, E., Karacam, H. and Bcer, V. (2007). Acute toxicity of carbaryl, methiocarb and carbosulfan to the rainbow trout (Oncorhynchus mykiss) and guppy (Poecilia reticulate), Turk. J. Vet. Anim. Sci., 31.
16. Maharaj S. (2005). Modelling the Behaviour and Fate of Priority Pesticides in South Africa. MSc Thesis, Department of Earth Sciences, University of Western Cape, South Africa.
17. Storrs, S. I. and Kiesiecker, (2004). Survivorship patterns of larval amphibians exposed to low concentrations of atrazine, Environmental Health Perspectives, 112 (10), 1054-105.
18. Scholz, N.L., Incardona, J.P., Baldwin, D.H., Berejikan, B.A., Dittman, A.H., Feist, B.E. and Jordan, C. (2003). Evaluating the sublethal impacts of current use pesticides on the environmental health of salmonids in Columbia River Basin. Bonneville Power Administration FY 2003 Provincial Project Review, 1-41.
19. Khan, M.Z., Tabassum, R., Naqvi, S.N.H., Shah, E.Z., Tabassum, F., Ahmad, I., Fatima, F. and Khan, M.F. (2003). Effect of cypermethrin and permethrin on cholinesterase activity and protein contents in

Rana tigrina (Amphibia), Turkey Journal of Zoology, 27, 243-246.

20. Radosevich, S.R., Holt J.S. and Ghersa C.M. (2007). Ecology of Weeds and Invasive Plants: Relationship to Agriculture and Natural Resource Management (3rd edn.), Wiley-Interscience, Hoboken, USA.
21. Wilson, C. (2009). Aquatic toxicology notes: predicting the fate and effects of aquatic and ditchbank herbicides. Soil and Water Science Department, Institute of Food and Agricultural Sciences, University of Florida, Gainesville, USA.
22. Pérez, G.L., Vera, M.S. and Miranda, L. (2011). Effects of Herbicide Glyphosate and Glyphosate-Based Formulations on Aquatic Ecosystems, Herbicides and Environment, Andreas Kortekamp (Ed.), ISBN: 978-953-307-476-4, InTech. http://www.intechopen.com/articles/show/title/effects-of-herbicide-glyphosate-and-glyphosate-based-formulations-on-aquatic-ecosystems (Accessed November, 2011).
23. Tsui, M.T.K. and Chu, L.M. (2003). Aquatic toxicity of glyphosate-based formulations: comparison between different organisms and the effects of environmental factors. Chemosphere, 52, 1189-1197.
24. Zhou, D-M., Wang, Y-J., Cang, L., Hao, X-Z., and Luo, X-S. (2004). Adsorption and cosorption of cadmium and glyphosate on two soils with different characteristics. Chemosphere, 57, 1237–1244.
25. Tu, M., Hurd, C. and Randall, J.M. (2001).Weed Control Methods Handbook: Tools and Techniques for Use in Natural Areas, The Nature Conservancy. http://tncweeds.ucdavis.edu (Accessed July, 2009).
26. WHO (World Health Organization) (1994). Environmental Health Criteria 159 – Glyphosate, International Programme on Chemical Safety. World Health Organization, Geneva, Switzerland.
27. Miller, A., Gervais, J.A., Luukinen, B., Buhl, K. and Stone, D. (2010). Glyphosate Technical Fact Sheet; National Pesticide Information Center, Oregon State University Extension Services. http://npic.orst.edu/factsheets/glyphotech.html (accessed January 2012).
28. Nordby, D.E., Hager, A.G. (2011). Herbicide formulations and calculations: active ingredient or acid equivalent, a Weed Fact sheet.

Integrated Pest Management Handbook, University of Illinois, USA. http://ipm.illinois.edu/weeds/aeai.pdf. Accessed 4 November 2011.

29. Stenersen, J. (2004). Chemical Pesticides: Mode of Action and Toxicology. CRC Press, Boca Raton, Florida, USA, 296 pp.

30. Schuette, J. (1998). Environmental fate of glyphosate, Environmental Monitoring & Pest Management, Department of Pesticide Regulation, Sacramento, CA 95824-5624, USA.

31. Meyer, M.T., Loftin, K.A., Lee, E.A., Hinshaw, G.H., Dietze, J.E. and Scribner, E.A. (2009). Determination of Glyphosate, its Degradation Product Aminomethylphosphonic Acid, and Glufosinate, in Water by Isotope Dilution and Online Solid-Phase Extraction and Liquid Chromatography/Tandem Mass Spectrometry: U.S. Geological Survey Techniques and Methods, book 5, chap. A10, 32p.

32. Herrmann, K.M and Weaver, L.M. (1999). The shikimate pathway. Annual Review of Plant Physiology and Plant Molecular Biology, 50, 473-503.

33. Giesy, J.P., Dobson, S. and Solomon, K.R. (2000). Ecotoxicological risk assessment for Roundup herbicide. Review of Environmental Contamination and Toxicology, 167, 35-120.

34. Glusczak L., Miron D. S., Moraes B. S., Simões R. R., Schetinger M. R. C., Morsch V. M. and Loro V. L. (2007). Acute effects of glyphosate herbicide on metabolic and enzymatic parameters of silver catfish (Rhamdia quelen). Comparative Biochemistry and Physiology-Part C, 146, 519-524.

35. El-Shebly A. A. and El-Kady M. A. H. (2008). Effects of glyphosate herbicide on serum growth hormone (GH) levels and muscle protein content in Nile Tilapia (Oreochromis Niloticus L.). Research Journal of Fisheries and Hydrobiology, 3(2), 84-88.

36. Rubach, M.N., Crum, S.J.H. and Van den Brink, P.J. (2011). Variability in the dynamics of mortality and immobility responses of freshwater arthropods exposed to Chlorpyrifos. Archives of Environmental Contamination and Toxicology, 60, 708-721.

37. Wirth, E.F., Lund, S.A. and Fulton, M.H. (2001). Determination of acute mortality in adults and sublethal embryo responses of Palaemonetes pugio to endosulfan and methoprene exposure. Aquatic Toxicology, 53, 9-18.

38. Key, P.B and Fulton, M.H. (2006). Correlation between 96-h mortality and 24 h acetylcholinesterase inhibition in three grass shrimp larval life stages. Ecotoxicology and Environmental Safety, 63, 389-392.

39. OECD (Organisation for Economic Co-operation and Development) (2005). Detailed review paper on aquatic arthropods in life-cycle and two-generation toxicity tests. An OECD Environment Health and Safety Publications Series on Testing and Assessment, Number 50. Environmental Directorate of Organization for Economic Co-operation and Development, Paris, France.

40. USEPA (United States Environmental Protection Agency) (2000). Methods for Measuring the Toxicity and Bioaccumulation of Sediment-Associated Contaminants with Freshwater Invertebrates. EPA/600/R-99/064. United States Environmental Protection Agency, Office of Prevention, Pesticides and Toxic Substances. Washington, USA.

41. Langdon, C.J., Harmon, V.L., Vance, M.M., Kreeger, K.E., Kreeger, D.A. and Chapman, G.A. (1996). A 7-d Toxicity test for marine pollutants using the Pacific mysid Mysidopsis intii. 1. Culture and protocol development. Environmental Toxicology and Chemistry, 15(10), 1815-1823.

42. Winkler, G. and Greve, W. (2002). Laboratory studies of the effect of temperature on growth, moulting and reproduction in the co-occurring mysids Neomysis integer and Praunus flexuosus. Marine Ecology Progress Series, 235, 177-188.

43. Verslycke, T.A., Fockedey, N., McKenney Jr, C.L., Roast, S.D., Jones, M.B., Mees, J. and Janssen, C.R. (2004). Mysid crustaceans as potential test organisms for evaluation of environmental endocrine disruption: a review. Environmental Toxicology and Chemistry, 23(5), 1219-1234.

44. Widarto, T.H., Krogh, P.H. and Forbes, V.E. (2007). Nonylphenol stimulates fecundity but not population growth rate (λ) of Folsomia candida. Ecotoxicology and Environmental Safety, 67, 369-377.

45. Lachaise, F., le Roux, A., Hubert, M. and Lafont, R. (1993). The moulting gland of crustaceans: localization, activity, and endocrine control (A Review). Journal of Crustacean Biology, 13(2), 198-234.

46. Spaziani, E., Mattson, M.P., Wang, W.L. and McDougall, H.E. (1999). Signaling pathways for ecdysteroid hormone synthesis in crustacean Y-organs. American Zoology, 39, 496-512.
47. Gorokhova, E. (2002). Moult cycle and its chronology in Mysis mixta and Neomysis integer (Crustacea, Mysidacea): implications for growth assessment. Journal of Experimental Marine Biology and Ecology, 278(2), 179-194.
48. Mu, X. and LeBlanc, G.A. (2002). Environmental antiecdysteroids alter embryo development in the crustacean Daphnia magna. Journal of Experimental Zoology, 292, 287-292.
49. Ketse, N. (2006). The effects of selected reference toxicants on embryonic development of the freshwater shrimp Caridina nilotica (Decapoda: Atyidae), MSc. Thesis, Unilever Centre for Environmental Water Quality (UCEWQ), Rhodes University, Grahamstown, South Africa.
50. Wester, P.W., van der Ven, L.T.M., Vethaak, A.D., Grinwis, G.C.M. and Vos, J.G. (2002). Aquatic toxicology: opportunities for enhancement through histopathology. Environmental Toxicology and Pharmacology, 11, 289-295.
51. Parvez, S. and Raisuddin, S. (2005). Protein carbonyls: novel biomarkers of exposure to oxidative stress-inducing pesticides in freshwater fish Channa punctata (Bloch). Environmental Toxicology and Pharmacology, 20, 112-117.
52. Richardson N., Gordon A.K., Muller W.J., Pletschke B.I. and Whitfield, A.K. (2010). The use of liver histopathology, lipid peroxidation and acetylcholinesterase assays as biomarkers of contaminant induced stress in the Cape stumpnose, Rhabdosargus holubi (Teleostei: Sparidae), from selected South African estuaries. Water SA, 36(4), 407-415.
53. Bernal-Hernandez, Y.Y., Medina-Dıaz, I.M., Robledo-Marenco, M.L., Velazquez-Fernandez, J.B., Giron-Perez, M.I., Ortega-Cervantes, L., Maldonado-Vazquez, W.A. and Rojas-Garcıa, A.E. (2010). Acetylcholinesterase and metallothionein in oysters (Crassostrea corteziensis) from a subtropical Mexican Pacific estuary. Ecotoxicology, 19, 819-825.

54. Kavitha, P.J. and Venkateswara, R. (2008). Toxic effects of chlorpyrifos on antioxidant enzymes and target enzyme acetylcholinesterase interaction in mosquito fish, Gambusia affinis. Environmental Toxicology and Pharmacology, 26, 192-198.

55. Modesto, K.A. and Martinez, C.B.R. (2010). Effects of Roundup Transorb on fish: hematology, antioxidant defenses and acetylcholinesterase activity. Chemosphere, 81, 781-787.

56. Menendez-Helman, R.A., Ferreyroa, G.V., dos Santos Afonso, M. and Salibian, A. (2012). Glyphosate as an Acetylcholinesterase Inhibitor in Cnesterodon decemmaculatus. Bulletin of Environmental Contamination and Toxicology, 88, 6-9.

57. Banerjee, B.D., Seth, V., Bhattacharya, A., Pasha, S.T. and Chakraborty, A.K. (1999). Biochemical effects of some pesticides on lipid peroxidation and free-radical scavengers. Toxicology Letters, 107, 33-47.

58. Barata, C., Varo, I., Navarro, J.C., Arun, S. and Porte, C. (2005). Antioxidant enzyme activities and lipid peroxidation in the freshwater cladoceran Daphnia magna exposed to redox cycling compounds. Comparative Biochemistry and Physiology, Part C, 140, 175-186.

59. Beuret, C.J., Zirulnik, F., and Gimenez, M.S. (2005). Effect of the herbicide glyphosate on liver lipoperoxidation in pregnant rats and their fetuses. Reproductive Toxicology, 19, 501–504.

60. Bold, T. (2007). Working for Water management treatment summary guide for aquatic and terrestrial alien plants, updated in May 2007, http://www.dwaf.gov.za/wfw/Control (Accessed July, 2009).

61. Romero D.M., Molina M.C.R. and Juarez A.B. (2011). Oxidative stress induced by a commercial glyphosate formulation in a tolerant strain of Chlorella kessleri. Ecotoxicology and Environmental Safety, 74, 741-747.

62. Mensah, P.K., Muller, W.J., Palmer, C.G., 2011. Acute toxicity of Roundup® herbicide to three life stages of the freshwater shrimp Caridina nilotica (Decapoda: Atyidae). Physics and Chemistry of the Earth, 36, 905-909.

63. Mensah, P.K., Muller, W.J., Palmer, C.G., 2012. Using growth measures in the freshwater shrimp Caridina nilotica as biomarkers of Roundup® pollution of South African freshwater systems. Physics and Chemistry of the Earth, 50-52, 262-268.

64. Mensah, P.K., Muller, W.J., Palmer, C.G., 2012. Acetylcholinesterase activity in the freshwater shrimp Caridina nilotica (Decapoda: Atyidae) as a biomarker of the herbicide Roundup® pollution of freshwater systems in South Africa. Water Science and Technology, 66(2), 402-408.

65. Mensah, P.K., Palmer, C.G., Muller, W.J., 2012. Lipid peroxidation in the freshwater shrimp Caridina nilotica as a biomarker of Roundup® herbicide pollution of freshwater systems in South Africa. Water Science and Technology, 65(9), 1660-1666.

7

CHAPTER

POTENTIAL HEALTH EFFECTS OF PESTICIDES

INTRODUCTION

In order for all pesticides to be effective against pests it should act in a way that are intended to control, they must be biologically active with analysis using antimicrobial test, toxicity level, and lethal dosage acceptable to be aware of the right dose for its application. Hence, pesticide has the ability to inhibit and control sthe metabolism of the target insect they are also potentially hazardous to humans, animals, other organisms, and the environment. In this connection, people who apply pesticides in any form or having a regular contact with pesticide must understand the extent of its lethal concentration, its toxicity, potential health effects, and preventative measures to reduce exposure to the products they use.

PESTICIDE TOXICITY AND EXPOSURE

The risk in using pesticides might lead to some degree of hazard, or the degree of danger involved in using a pesticide under a given set of conditions. Hazard depends on the toxicity on the application of the pesticide as well as the amount of pesticide applied and is often illustrated with the following equation:

Hazard = Toxicity x Exposure

The level of lethality of the applied pesticide can be measured with its capacity or ability to hazardous effect. The toxicity of a particular pesticide is determined by subjecting test animals to varying experimental analysis such as lethal dosage LD_{50} and at the same time, the lethal concentration LC_{50} present in the bioactive composition as the active ingredient and each of its formulated products. Hence, it is the active ingredient is the chemical component in the pesticide product that controls the activity of the tested pest. By understanding the level of concentration of its efficacy as pesticides, a user can minimize the potential hazard by selecting the pesticide with the lowest toxicity that will control the pest.

Applicators may have little or no control over the availability of low-toxicity products or the toxicity of specific formulated products. However, applicators can minimize or nearly eliminate exposure and thus reduce hazard by following the label instructions, using personal protective clothing and equipment (PPE), and handling the pesticide properly. For example, more than 95 percent of all pesticide exposures come from dermal exposure, primarily to the hands and forearms. By wearing a pair of unlined, chemical-resistant gloves, this type of exposure can be nearly eliminated.

Acute Toxicity and Acute Effects

Acute toxicity of a pesticide refers to the chemical's ability to cause injury to a person or animal from a single exposure, generally of short duration. The harmful effects that occur from a single exposure by any route of entry are termed "acute effects." The four routes of exposure are dermal (skin), inhalation (lungs), oral (mouth), and the eyes. Acute toxicity is determined by examining the dermal toxicity, inhalation toxicity, and oral toxicity of test animals. In addition, eye and skin irritation are also examined.

Acute toxicity is measured as the amount or concentration of a toxicant—the a.i.—required to kill 50 percent of the animals in a test population. This measure is usually expressed as the LD_{50} (lethal dose 50) or the LC_{50} (lethal concentration 50). Additionally, the LD_{50} and LC_{50} values are based on a single dosage and are recorded in milligrams of pesticide per kilogram of body weight (mg/kg) of the test animal or in parts per million (ppm). LD_{50} and LC_{50} values are useful in comparing the toxicities of different active ingredients and different formulations containing the same active ingredient. The lower the LD_{50} or LC_{50} value of a pesticide product, the greater its toxicity to humans and animals. Pesticides with a high LD_{50} are the least toxic to humans if used according to the directions on the product label.

Chronic Toxicity and Chronic Effects

The acute effect on the toxicity of a pesticide is determined by subjecting test animals to long-term exposure to the active ingredient. Any harmful effects that occur from small doses repeated over a period of time are termed "chronic effects." Suspected chronic effects from exposure to certain pesticides include birth defects, toxicity to a fetus, and production of benign or malignant tumors, genetic changes, blood disorders, nerve disorders, endocrine disruption, and reproduction effects. The chronic toxicity of a pesticide is more difficult than acute toxicity to determine through laboratory analysis.

Pesticide Signal Words

Pesticides being invented for use can be categorized on the basis of their relative acute toxicity that can be detected in their LD_{50} or LC_{50} values. Pesticides that are classified into three ways , highly toxic, moderately toxic, slightly toxic and non toxic depending on its effect either oral, dermal, or inhalation toxicity must have the signal words. A proper label should be added as DANGER and POISON with a skull or crossbone incapsulated in every label that should be openly and prominently displayed on the front panel of the package label to signify its harm. But before hand, this should undergo a series of experimental analysis to know its toxicity. The acute (single dosage) oral LD_{50} for pesticide products in this group ranges from a trace amount to 50 mg/kg. For example, exposure of a few drops of a material taken orally could be fatal to a 150-pound person.

Some pesticide products have just the signal word DANGER, which tells you nothing about the acute toxicity, just that the product can cause severe eye damage or severe skin irritation.Pesticide products considered moderately toxic (Toxicity Category II) must have the signal word WARNING sign displayed on the product label. In this category, the acute oral LD_{50} ranges from 50 to 500 mg/kg. A teaspoon to an ounce of this material could be fatal to a 150-pound person.

Pesticide products classified as either highly toxic, moderately toxic, slightly toxic and non toxic are required to have the signal word CAUTION on the pesticide label. Acute oral LD_{50} values in this group are greater than 500 mg/kg. An ounce or more of this material could be fatal to a 150-pound person.

Table 7. 1 summarizes the LD_{50} and LC_{50} values for each route of exposure for the four toxicity categories and their associated signal word. For example, an active ingredient with a dermal LD_{50} of 1,000 mg/kg would be in Toxicity Categor`y II with a WARNING signal word. Keep in mind that an active ingredient may have a high LD_{50} placing it in a Toxicity Category II, III, or IV but also have corrosive eye/skin effects that take priority and place it in Toxicity Category I.

Table 7.1: Summary of the Toxicity levels due to exposure to pesticide.

Routes of Exposure	Toxicity Category			
	I	II	III	IV
Adapted from 40 CFR Part 156.				
Oral LD50	Up to and including 50 mg/kg	50–500 mg/kg	500–5,000 mg/kg	>5,000 mg/kg
Inhalation LC50	Up to and including 0.2 mg/l	0.2–2 mg/l	2–20 mg/l	>20 mg/l
Dermal LD50	Up to and including 200 mg/kg	200–2,000 mg/kg	2,000–20,000 mg/kg	>20,000 mg/kg
Eye Effects	Corrosive corneal opacity not reversible within 7 days	Corneal opacity reversible within 7 days; irritation persisting for 7 days	No corneal opacity; irritation reversible within 7 days	No irritation
Skin Effects	Corrosive	Severe irritation at 72 hours	Moderate irritation at 72 hours	Mild or slight irritation at 72 hours
Signal Word	DANGER POISON	WARNING	CAUTION	CAUTION

SYMPTOMS OF PESTICIDE POISONING

The symptoms of pesticide poisoning can range from a mild skin irritation to coma or even death. Different classes or families of chemicals cause different types of symptoms. Individuals also vary in their sensitivity to different levels of these chemicals. Some people may show no reaction to an exposure that may cause severe illness in others. Because of potential health concerns, pesticide users and handlers must recognize the common signs and symptoms of pesticide poisoning.

The effects, or symptoms, of pesticide poisoning can be broadly defined as either topical or systemic. Topical effects generally develop at the site of pesticide contact and are a result of either the pesticide's irritant properties which can either the active and/or inert ingredient or an allergic response by the victim. Dermatitis, or inflammation of the skin, is accepted as the most commonly reported topical effect associated with pesticide exposure. Symptoms of dermatitis range from reddening of the skin to rashes and/or blisters.

Some individuals tend to cough, wheeze, or sneeze when exposed to pesticide sprays. Some individuals react to the strong odor and irritating effects of petroleum distillates used as carriers in pesticide products. One symptom is that the eyes, mucous membranes of the nose, and even the sensitive linings of the mouth and back of the throat feel raw and scratchy. This symptom usually subsides within a few minutes after a person is removed from the exposure to the irritant. However, a reaction to a pesticide product that causes someone not only to sneeze and cough but also to develop severe acute respiratory symptoms is more likely to be a true hypersensitivity or allergic reaction. Symptoms of a true allergic reaction range from reddening and itching of the eyes and skin to respiratory discomfort often resembling an asthmatic condition.

Systemic effects are quite different from topical effects. They often occur away from the original point of contact as a result of the pesticide being absorbed into and distributed throughout the body. Systemic effects often include nausea, vomiting, fatigue, headache, and intestinal disorders. In advanced poisoning cases, the individual may experience changes in heart rate, difficulty breathing, convulsions, and coma, which could lead to death.

Seeking Medical Attention

Be alert for the early signs and symptoms of pesticide poisoning in yourself and others. These often occur immediately after exposure, but they could be delayed for up to 24 hours. If you are having symptoms but are unsure if they are pesticide related, at least notify someone in case your symptoms become worse. But when symptoms appear after contact with pesticides, you should seek medical attention immediately.

If safe to do so, take the pesticide container to the telephone. However, if the pesticide container is contaminated, write down the product name, active ingredient(s) and percentage, and the EPA registration number. The product label provides medical personnel information such as active ingredients, an antidote, and an emergency contact number for the manufacturer. If the Material Safety Data Sheet is available, take this also because it contains additional information for medical personnel.

If you must go to the hospital or doctor's office, take the entire pesticide container, including the label, with you. In order to avoid inhaling fumes or spilling the contents, make sure the container is tightly sealed and place into a plastic bag if possible. The pesticide container should never be placed in the enclosed passenger section of your vehicle.

HARMFUL EFFECTS OF SOME PESTICIDE FAMILIES

Fungicides

The acute toxicity of fungicides to humans is generally considered to be low, but fungicides can be irritating to the skin and eyes. Inhalation of spray mist or dust from these pesticides may cause throat irritation, sneezing, and coughing. Chronic exposures to lower concentrations of fungicides can cause adverse health effects. Most cases of human fungicide poisonings have been from consumption of seed grain. To prevent these types of poisonings, fungicide treatment now includes a brightly colored dye to clearly indicate that the seed has been treated. Table 7.2 summarizes the signs and symptoms of acute exposures to commonly used fungicides.

Table 7.2: Signs and symptoms of acute exposure with fungicide active ingredients.

Active Ingredient	**Brand Name**	**Signs and Symptoms**
Azoxystrobin	Abound, Quadris	Irritating to skin, eyes, respiratory tract
Captan	Captol, Ortho-cide	Irritating to skin, eyes, respiratory tract
Chlorothalonil	Bravo, Daconil	Irritation to skin, mucous membranes of the eye, respiratory tract. Allergic contact dermatitis.
Copper Compounds	Bordeaux mixture, Copper sulfate	Irritating to skin, eyes, respiratory tract. Salts are corrosive to mucous membranes and cornea Metallic taste, nausea, vomiting, intestinal pain.
Mancozeb	Dithane M-45, Manzate 200	Irritating to skin, eyes, respiratory tract
Maneb	Dithane M-22, Manzate	Irritating to skin, eyes, respiratory tract. Skin disease in occupationally exposed individuals.
Pentachloronitrobenzene	PCNB, Terraclor	Allergic reactions
Sulfur	Cosan, Thiolux	Irritating to skin, eyes, respiratory tract. Breath odor of rotten eggs. Diarrhea. .Irritant dermatitis in occupationally exposed individuals
Thiram	Polyram-Ultra, Spotretc-F	Irritating to skin, eyes, respiratory mucous membranes
Ziram	Cuman, Vancide	Irritating to skin, eyes, respiratory tract Prolonged inhalation causes neural and visual disturbances

Herbicides

In general, herbicides have a low acute toxicity to humans because the physiology of plants is so different than that of humans. However, there are exceptions; many can be dermal irritants since they are often strong acids, amines, esters, and phenols. Inhalation of spray mist may cause coughing and a burning sensation in the nasal passages and chest. Prolonged inhalation sometimes causes dizziness. Ingestion will usually cause vomiting, a burning sensation in the stomach, diarrhea, and muscle twitching. Table 7.3 summarizes the signs and symptoms of acute exposures to commonly used herbicides.

Table 7.3: Signs and symptoms of acute exposure with herbicides active ingredients.

Active Ingredient	**Brand Name**	**Signs and Symptoms**
2,4-dichlorophenoxy-acetic acid	2,4-D, Barrage	Irritating to skin, mucous membranes. Vomiting, headache, diarrhea, confusion. Bizarre or aggressive behavior. Muscle weakness in occupationally exposed individuals.
Acetochlor	Harness, Surpass	Irritating to skin, eyes, respiratory tract.
Atrazine	Aatrex, Atranex	Irritating to skin, eyes, respiratory tract. Abdominal pain, diarrhea, vomiting. Eye irritation, irritation of mucous membranes, skin reactions.
Dicamba	Banvel, Metambane	Irritating to skin, respiratory tract. Loss of appetite (anorexia), vomiting, muscle weakness, slowed heart rate, shortness of breath. Central nervous system effects.
Glyphosate	Rodeo, Roundup	Irritating to skin, eyes, respiratory tract.
Mecoprop	Kilporp, MCPP	Irritating to skin, mucous membranes. Vomiting, headache, diarrhea, confusion. Bizarre or aggressive behavior. Muscle weakness in occupationally exposed individuals.
Metolachlor	Bicep, Dual	Irritating to skin, eyes
Paraquat	Gramoxone	Burning in mouth, throat, chest, upper abdomen. Diarrhea. Giddiness, headache, fever, lethargy. Dry, cracked hands, ulceration of skin.
Pendimethalin	Prowl, Stomp	Irritating to skin, eyes, respiratory tract
Propanil	Propanex, Stampede	Irritating to skin, eyes, respiratory tract

Insecticides

Insecticides cause the greatest number of pesticide poisonings in the United States. The most serious pesticide poisonings usually result from acute exposure to organophosphate and carbamate insecticides. Organophosphate insecticides include chlorpyrifos, diazinon, dimethoate, disulfoton, malathion, methyl parathion, and ethyl parathion. The carbamate compounds include carbaryl, carbofuran, methomyl, and oxamyl. Organophosphates and carbamates inhibit the enzyme cholinesterase, causing a disruption of the nervous system. All life forms with cholinesterase in their nervous system, such as insects, fish, birds, humans, and other mammals, can be poisoned by these chemicals.

Table 7. 4: summarizes the signs and symptoms from acute exposures to commonly used insecticides.

Active Ingredient	**Brand Name**	**Signs and Symptoms**
Acephate (organophosphate)	Orthene	Headache, excessive salivation and tearing, muscle twitching, nausea, diarrhea. Respiratory depression, seizures, loss of consciousness. Pinpoint pupils.
Aldicarb (N-methyl carbamate)	Temik	Malaise, muscle weakness, dizziness, sweating. Headache, salivation, nausea, vomiting, abdominal pain, diarrhea. Nervous system depression, pulmonary edema in serious cases.
Carbaryl (N-methyl carbamate)	Sevin	Malaise, muscle weakness, dizziness, sweating. Headache, salivation, nausea, vomiting, abdominal pain, diarrhea. Nervous system depression, pulmonary edema in serious cases.
Chlorpyrifos (organophosphate)	Dursban	Headache, excessive salivation and tearing, muscle twitching, nausea, diarrhea. Respiratory depression, seizures, loss of consciousness. Pinpoint pupils.
Endosulfan (organochlorine)	Thiodan	Itching, burning, tingling of skin. Headache, dizziness, nausea, vomiting, lack of coordination, tremor, mental confusion. Seizures, respiratory depression, coma.
Malathion (organophosphate)	Cythion	Headache, excessive salivation and tearing, muscle twitching, nausea, diarrhea. Respiratory depression, seizures, loss of consciousness. Pinpoint pupils.
Methyl Parathion (organophosphate)	Penncap-M	Headache, excessive salivation and tearing, muscle twitching, nausea, diarrhea. Respiratory depression, seizures, loss of consciousness. Pinpoint pupils.
Phosmet (organophosphate)	Imidan	Headache, excessive salivation and tearing, muscle twitching, nausea, diarrhea. Respiratory depression, seizures, loss of consciousness. Pinpoint pupils.
Pyrethrins (natural origin)		Irritating to skin and upper respiratory tract. Contact dermatitis and allergic reactions—asthma.

Pyrethroids (synthetic pyrethrin)	Cypermethrin, permethrin	Abnormal facial sensation, dizziness, salivation, headache, fatigue, vomiting, diarrhea. Irritability to sounds or touch. Seizures, numbness.

To understand how the organophosphate and carbamate insecticides affect the nervous system, one needs to understand how the nervous system actually works. The nervous system, which includes the brain, is the most complex system in the body consist- ing of millions of cells that make up a communications system within the organism. Messages or electrical impulses (stimuli) travel along this complex network of cells. Nerve cells or neurons do not physically touch each other; rather there is a gap or synapse between cells. The impulses must cross or "bridge" the synapse between nerve cells in order to keep the message moving along the entire network.

When an impulse reaches the synapse, the chemical acetylcholine is released to carry the message on to the next cell. Acetylcholine is the primary chemical responsible for the transmission of nerve impulses across the synapse of two neurons. After the impulse is transmitted across the synapse, the acetylcholine is broken down by the enzyme cholinesterase. Once this occurs, the synapse is "cleared" and ready to receive a new transmission.

Organophosphate and carbamate insecticides inhibit the activity of cholinesterase, resulting in a buildup of acetylcholine in the body. An increase in acetylcholine results in the uncontrolled flow of nerve transmissions between nerve cells. The nervous system becomes "poisoned"; the accumulation of acetylcholine causes the continual transmission of impulses across the synapses.

The effects of organophosphate or carbamate poisoning can result in both systemic and topical symptoms. Direct exposure of the eye, for example, can cause topical symptoms such as constriction of the pupils, blurry vision, an eyebrow headache, and severe irritation and reddening of the eyes. Symptoms and signs of systemic poisonings are almost entirely due to the accumulation of acetylcholine at the nerve endings.

The onset of symptoms depends on the route of entry and the severity of the poisoning. Gastric symptoms such as stomach cramps, nausea, vomiting, and diarrhea appear early if the material has been ingested. Similarly, salivation, headache, dizziness, and excessive secretions that cause breathing difficulties are initial symptoms if the material has been inhaled. Involvement of the respiratory muscles can result in respiratory failure. Stomach, intestinal, and respiratory symptoms usually appear at the same time if the pesticide is absorbed through the skin. In children, the first symptom of poisoning may be a convulsion.

In advanced poisonings, the victim is pale, sweating, and frothing at the mouth. The pupils are constricted and unresponsive to light. Other symptoms include changes in heart rate, muscle weakness, mental confusion, convulsions, and/or coma. The victim may die if not treated.

Cholinesterase Testing

Those who regularly work with organophosphates and carbamates should consider having periodic cholinesterase tests. The blood cholinesterase test measures the effect of exposure to organophosphate and carbamate insecticides. Since cholinesterase levels can vary considerably among individuals, a "baseline" must be established for each person. In fact, a small percentage of the population has a genetically determined low level of cholinesterase. Even minimal exposure to cholinesterase inhibitors can present a substantial risk to these people. Baseline testing should always be done during the time of year when pesticides are not being used, or at least 30 days from the most recent exposure. Establishing a baseline value often requires two tests performed at least 72 hours apart but within 14 days of each other. If the test results differ by as much as 20 percent, a third test is often recommended.

Cholinesterase tests can be repeated during times when organophosphate and carbamate insecticides are being used and then compared with the baseline level. The purpose of routine cholinesterase monitoring is to enable a physician to recognize the occurrence of excessive exposure to organophosphates and carbamates. If a laboratory test shows a cholinesterase drop of 30 percent below the established baseline, the worker should be retested immediately. If a second test confirms the drop in cholinesterase, the pesticide handler or agricultural worker should be removed from further contact with organophosphate and carbamate insecticides until cholinesterase levels return to the preexposure baseline range. Your primary care physician can help to establish the frequency of this testing program.

EXPOSURE AND PREVENTATIVE MEASURES

As mentioned earlier, the hazard or risk involved with using a pesticide depends on both the toxicity of the product and the amount of exposure to the product (Hazard = Toxicity x Exposure). Ideally, use a low-toxicity product when possible, but even they can be harmful if your exposure level is high. However, regardless of the product's toxicity, if the exposure level is low, then the hazard will also be low. To reduce the possibility of exposure and to protect your health, always wear the personal protective equipment (PPE) as indicated on the product label. The following are general PPE guidelines to protect against the four routes of entry.

Dermal

More than 95 percent of all exposures are dermal. Dermal absorption may occur as the result of a splash, spill, or drift or when cleaning or repairing equipment. Wear unlined, chemical-resistant gloves to eliminate most dermal exposures. Minimum dermal protection for most pesticides consists of a long-sleeved shirt, long trousers, gloves, and proper footwear. For extra precaution, consider wearing coveralls, a waterproof hat, and unlined rubber boots. Additionally, wearing a liquid- proof apron or rain suit is recommended when mixing and pouring concentrates or when using highly toxic products.

Inhalation

For many toxic chemicals, the respiratory (breathing) system is the quickest and most direct route of entry into the circulatory system. Respiratory protection is especially important when pesticide powders, dusts, gases, vapors, or small spray droplets can be inhaled.

Use the respirator as designed for its intended use, and always follow the manufacturer's instructions. Select only equipment approved by the National Institute of Occupational Safety and Health (NIOSH) and the Mine Safety and Health Administration (MSHA).

Oral

Accidental oral exposure most frequently occurs when pesticides have been taken from the original container and put into an unlabeled bottle or food container. Unfortunately, children are the most common victims in these situations. Store pesticides only in their original containers, and keep the original label attached to the container. Store pesticides only in their original containers and keep the original label attached to the container. Store in a locked cabinet and/or on a high shelf to keep out of the reach of children. Never use your mouth to clear a spray line or to siphon a pesticide from a tank or container. After handling or working with pesticides, wash your hands and face thoroughly with soap and water before eating, drinking, or smoking.

Eyes

Eyes are very sensitive to many pesticides and, considering their size, are able to absorb large amounts of chemical. Serious eye exposure can result from a splash, spill, or drift or by rubbing the eyes with contaminated hands or clothing. Tight-fitting chemical splash goggles or a full-face shield should be worn if there is any chance of getting pesticides in the eyes, especially when pouring or mixing concentrates and handling dusts. When pouring from a container, keep the container below eye level to avoid splashing or spilling chemicals on your face or protective clothing.

All pesticides have the potential to be harmful to humans, animals, other living organisms, and the environment if used incorrectly. The key to reducing health hazards when using pesticides is to always limit your exposure by wearing PPE and use a low-toxicity pesticide when available. Reading the label and practicing safe work habits will minimize hazards from the use of pesticides.

REFERENCES

1. Aspelin AL, Grube AH (1998) Pesticide Industry Sales and Usage: 1996 and 1997 Market Estimates. Office of Prevention, Pesticides and Toxic Substances, U. S. Environmental Protection Agency. 733-R-98- 0001. Washington, DC 20460. 37 pp.
2. Carpenter MJ, Ware GW (2004) Defending Pesticides in Litigation, 14th Ed. West-Thomson. St. Paul, MN 55123. 763 pp.
3. Charles D. 2001. Lords of the harvest: Biotech, big money, and the future of food. Perseus Publishing, Cambridge, MA, 348 pp.

4. Crop Protection Handbook (2004) Meister Media Worldwide, Willoughby, Ohio. Vol. 90, 900 pp.
5. EXTOXNET: Extension Toxicology Network. Oregon State University. Accessed September 20, 2006.
6. Mayer MS, McLaughlin JR (1990) Insect pheromones and sex attractants. CRC Press, Boca Raton, FL. 235 pp.
7. Reigart, J. R., and J. R. Roberts. Recognition and Management of Pesticide Poisoning. 5th ed. U.S. EPA Office of Pesticide Programs.
8. Salgado VL (1997) The modes of action of spinosad and other insect control products. Down to Earth 52(2):35-43. Dow AgroSciences, Midland, MI.
9. Thomson WT (1999-2000 Revision). Agricultural Chemicals, Book III, Miscellaneous Agricultural Chemicals. Thomson Publications, Fresno, California. 189 pp.
10. Thomson WT (2001) Agricultural Chemicals, Book I, Insecticides. Thomson Publications, Fresno, California. 249 pp.
11. Tomlin C (ed) (2000) The Pesticide Manual, 12th Ed. British Crop Protection Council. Blackwell Scientific Publications, Cambridge, Massachusetts. 1250 pp.
12. Ware GW, Whitacre DM (2004) The Pesticide Book, 6th Ed. 496 pp. Meister Media Worldwide, Willoughby, Ohio. (ISBN 1892829-11-8)

8
CHAPTER

ECOLOGICAL IMPACT OF ORGANOCHLORINE ENDOSULFAN

INTRODUCTION

Pesticide had been used in different areas such as agricultural production, industry, homecare and farmlots. This was noted toaffect and contaminate terrestrial and aquatic environments worldwide. Food and Authority has defined the term of pesticide, "As any compounds or mixture of compounds that its main function is for preventing, destroying or controlling any pest, including human or animal disease. Wherein sometimes unwanted species of plants or animals may be harmed unintentionally causing harm during its procedure or interferes in the production, processing, storage, transport or production of food, agricultural commodities, wood and wood products or animal feedstuffs, or substances which may be administered to animals for the control of insects, arachnids or other pests in or on their bodies. Some classes under pesticides are classified as herbicides, insecticides, fungicides, rodenticides, pediculocides and biocides.

Further pesticides can be grouped into different chemical functionalities. One of the features that make pesticide functions unique is that when applied to pests can develop resistance to its use thus, it needs to formulate another pesticide with another chemical compositions.

Another feature of pesticide is that it is dose persistent, it means that it will increase its dosage concentration of the pesticide can be used to counteract the resistance and increases also its chemical residues. An example of this pesticides that has dose persistent are those insecticides belonging to prominent families that include organochlorines, organophosphates and carbamates. However, each pesticide substance has a particular behavior in the environment.

Environmental peristence can be affected by different principalities such temperature, light, moisture content, humidity, alkalinity and acidity and finally its microbial interaction that can breakdown this compounds into minute particle. In many situations, the exact role of these factors and their interactions are not known. The effects of these substances on non target organisms, their accumulation in life process or their transfer to the higher trophic levels of the food-chain.

HISTORY OF PESTICIDES

Even before 2000 BC, humans have utilizes pesticides for crop protection and this was prevented for use when DDT was discovered with its negative effect. Especially when organochlorines one of the main component of DDT was being opposed by Rachel Carson. She found out that organochlorines are highly persistent in soil due to its low solubility. These was then replaced in U.S. by organophosphates and carbamates. Since, then pyrethrin compounds have become the dominant insecticides. The first generation of pesticides introduced over 40 years ago includes the Organochlorine compounds (OC) and Organophosporous (OP). Later they become unacceptably persistent in the environment, although at the time of their introduction, they seemed to be promising candidates. In 1940s manufacturers began to produce large amounts of synthetic pesticides and their use became widespread. Pesticide use has increased 50-folds and since 1950 and 2.3 million tons (2.5 million short tons) of industrial pesticides are now used every year. Seventy-five percent of all pesticides in the world are used in developed countries, but use in developing countries is in an increasing trend.

IMPACT OF ORGANOCHLORINE PESTICIDE

Chlorinated pesticides such as DDT, endosulfan, heptachlor, aldrin and another type of this chlorinated containing pesticide have been used primarily for the protection of agricultural products which sole function is for preventing airborne and viral disease that is cause by infiltration of this pest. These chlorinated pesticides persist in the environment over a long period of time which undergoes bioaccumulation and bio magnifications and can have a higher to lower lethal concentration to non-target organisms including human beings.

The manufacture of pesticide with chlorinated composition has been banned but highly industrialized countries but still being used in most of developing countries. Another thing is that pesticide influences negative impact in the environment which was started way back in 1960's, wherein it was discovered that DDT interferes in th reproduction of many aquatic life such as fishes and even birds which can be diversified to biodiversity. It was Rachel Carson who has written about this biological magnification in his best-selling book Silent Spring. The agricultural use of DDT is now banned under the Stockholm Convention on Persistent Organic Pollutants (POPs), but it is still used in some developing nations to prevent malaria and other tropical diseases by spraying on interior walls to kill or repel mosquitoes.

About 98% of the insecticide being sprayed and 95% of herbicides can penetrate the target and non-target species wherein some of its content can penetrate the in air, water and even in soil. A pesticide may drift in when pesticides suspended in the air for a long period of time and this particles are carried by air through the wind in different areas of the earth. Pesticides can also be a contributing factor as one of the causes to water pollution and the characteristics of pesticides that has resisting properties can also contribute ot soil contamination due to the presence of organic pollutants from pesticides.And a terrifying incident in Bhopal, India which had happened wherein about 40 tons of methyl isocyanate (MIC) gas was accidentally release which kills about 2,000 people who had affected by this incidents through inhalation and even ingestion from hence this type of pesticide can be synthesized into carbonate pesticides which high toxic when ingested and inhaled.

The use of pesticides for the control of crop pests and weeds has increased. In addition to this, discharges from the pesticide manufacturing plants, accidental spills and natural processes as dilution, surface run off and leaching are the cause of the occurrence of xenobiotic compounds in surface waters. The effects of pesticides on aquatic environments are also due to their degradation products, which can be more toxic than the original substances (Thurman et al., 1992). There is evidence that the OP compounds are sufficiently persistent to reach the marine environment at concentrations high enough to affect aquatic fauna and flora. They are toxic to invertebrates and vertebrate aquatic animals at very low concentrations (ppb or ppm level). Shrimp and fish are particularly sensitive. The presence of pesticides in costal environment is not entirely unexpected, since major agricultural areas are located in coastal plains and river valleys. However, occasionally massive fish and shrimp kills have been reported in estuarine and coastal lagoon ecosystems. Although the agents for this have not clearly identified, agrochemicals from adjacent agricultural fields have long been suspected as the lethal toxicants involved.

BIOMAGNIFICATIONS OF PESTICIDES

Even small concentrations of chemicals in the environment can find their way into organisms, accumulate in food chains leading to higher doses causing problems. Biomagnifications occurs when the pollutant is long-lived, mobile, soluble in fats and biologically active. If a pollutant is short-lived, chances for breaking down before it can become dangerous are more. If it is not mobile, it will stay in one place and is unlikely to be taken up by organisms. If a pollutant is soluble in water it will be excreted by the organism. Pollutants that dissolve in fats are however, retained for a long time in the organisms.

Biological magnification often refers to the process whereby certain substances such as pesticides or heavy metals move up in food chain, work their way into rivers or lakes and are eaten by aquatic organisms such as fish, which in turn areeaten by large birds, animals or humans. The substances become concentrated in tissues or internal organs as they move up the food chain. Pesticide degradation is a process involving three phases (Hatzios, 1991 and Shimabukuro, 1985). Phase I process involves oxidation, reduction, or hydrolysis, which changes the parent compound more water soluble and less toxic degradation product. Oxygenation is the most frequent first step in the biotransformation of pesticides.

Many of these reactions are mediated by oxidative enzymes viz. cytochromeP450s, peroxidases and polyphenol oxidases. The second phase involves conjugation of a pesticide or pesticide metabolites to a sugar, amino acid or glutathione, which increases the water solubility and reduces the toxicity compared to the parent compound. Generally, metabolites obtained from phase II have little or no toxicity, and may be stored in cellular organelles. Finally, the third phase involves conversion of phase II metabolites into secondary conjugates, which are generally nontoxic. Insecticides such as organochlorine hydrocarbons could be separated into dichlorodiphenyl ethanes, cyclodiene compounds and other related compounds.

Chlorinated pesticides persist in the environment for very long periods, undergo bioaccumulation and biomagnifications and therefore impart toxicity to non-target organisms including human beings. They operate by disrupting the sodium/potassium balance of the nerve fibers, forcing them to transmit continuously. Their toxicities vary greatly, but they have been phased out because of their persistence and potential to bioaccumulation. Accordingly, manufacture and the use of these chlorinated pesticides had either been banned or severely restricted in most of the developed and developing countries.

ENDOSULFAN

Endosulfan (6, 7,8,9,10,10 – hexachloro – 1, 5, 5a, 6, 9, 9a hexa hydro – 6, 9 – methano 2, 3, 4 benzo (e) dioxathiepen-3oxide) is a chlorinated hydrocarbon insecticide and acaricide of the cyclodiene subgroup which acts as a poison to a wide variety of insects and mites on contact. It is formed by the reaction of thionylchloride with thiodan. Endosulfan is a derivative of hexachlorocyclopentadiene and is chemically similar to aldrin, chlordane and heptachlor. Specifically, it is produced by the DielsAlder reaction of hexachlorocyclopentadiene with cis-butent-1, 4-diol and subsequent reaction of the adduct with thionyl chloride.

Technical endosulfan is a 7:3 mixture of stereoisomers, designated α andβ. α and β-endosulfans are conformational isomers arising from the pyramidal stereochemistry of sulfur. α-Endosulfan is the more thermodynamically stable of the two, thus β-endosulfan irreversibly converts to the α form, although the conversion is slow. These two isomers persist longer under more acidic conditions. The dissipation depends on a multitude of factors such volatilization, alkaline hydrolysis and photodecomposition, besides the presence of fertilizer, crop pattern, atmospheric temperature, rain and biotic conversions among others (Goebel et al., 1982). Endosulfan was introduced in early 1950s. In1954 Hoechst AG (now Bayer Crop Science) won USDA's approval of endosulfan in the US.

The Australian Pesticides and Veterinary Medicines Authority, the responsible federal authority declared that shunning evidence of the dangers is available regarding the use of endosulfan. The chemical has been banned worldwide by the Stockholm Convention of Persistent Organic Pollutants due to its unacceptable neurological and reproductive risks to farm works and wildlife. Many International studies identify endosulfan as an endocrine disruptor which interferes with hormones and reproduction. Endosulfan is subjected to long range atmospheric transport, i.e. it can travel long distances from where it is used. The endosulfan commonly contaminates air, water, plants and fish of national parks. Most of these parks are far from areas where endosulfan was used. Endosulfan had also been detected in dust from the Sahara Desert collected in the Caribbean after blown across the Atlantic Ocean (Ramnarine and Kristy, 2008).

ECOLOGICAL IMPACTS OF ENDOSULFAN

Endosulfan is extremely toxic to fish and aquatic invertebrates (Sunderam et al., 1992) and it has been implicated increasingly in gonadal toxicity (Sinha et al., 1997), genotixicity (Chaudhuri et al., 1999) and neurotixicity (Paul and Balasubramanian, 1997; Siddique et al., 2003) while provoking chronic symptoms like testicular and prostate cancer (Saiyed et al., 2003) breast cancer and sexual abnormality (Arnold et al., 1996) in numerous mammalian species. Persistence of endosulfan in soil and water environments had been widely reported by different researchers under different conditions (Rao and Murty, 1980; Guerin and Kennedy, 1992 andSethunathan et al., 2002). Because of its abundant usage and potential transport, endosulfan contamination is frequently found in the environment at considerable distances from the point of its original applications (Man singh and Wilson, 1995 and Miles and Pfeuffer, 1997).

DeLorenzo et al., (2002) reported on the bioconcentration factors of freshwater green algae (Pseudokirchnereilla subcapitatum) and freshwater water flea (Daphnia magna) as 2682 and 3278, respectively, with little evidence on bioaccumulation in phytoplankton and zooplankton. In contrast, endosulfan residues in farmed snakehead and carp fish (Channa striata and Catla catla), from a region in India with intensive agricultural activities, revealed higher bioaccumulation factors (BAFs) of log 4.5 (±0.45) (Amaraneni, 2002).

Endosulfan is classified as a priority pollutant by international agencies (Keith and Telliard, 1979). High doses of endosulfan in humans primarily affect the central nervous system, resulting in respiratory depression, dry mouth, convulsions and possibly death due to major biotransformation products like Endosulfan sulfate and Endosulfan diol are formed. Endosulfan sulfate is more toxic and more persistent compared to the parent Endosulfan (Kaur et al., 1998).

ENDOSULFAN TOXICITY

In an assessment of Endosulfan toxicity to a range of freshwater organisms representing various tropic levels in farm ditches in the Lower Fraser Valley of BC, Canada. Wan et al. (2005) found that endosulfan to be the more toxic of the two isomers. However, more importantly, a combination of the isomers and endosulfan sulfate appeared to be more potent than any single endosulfan isomer. The study concluded that the higher concentration of endosulfan used in Lower Fraser Valley farms drained into salmon which often exceeded the 96-h LC_{50} values of juvenile salmon. The concentration measured in ditch sediments were negatively affect the development and survival of benthic organisms. For biotic (microbial) degradation, Guerin (1999) found extensive degradation of endosulfan (N 85%) by indigenously mixed microbial populations under anaerobic conditions, but did not find significant enhancement of degradation from enriched inoculate. These results demonstrate the biodegradation potential of α-endosulfan in low oxygen containing environments such as sediments and this may have implications for interpreting historic accumulation trends of the α-isomer in freshwater sediments.

Endosulfan sulfate is likely to represent the predominant residue of endosulfan over time (Guerin, 2001). Aquatic half-lives have been reported to be respectively 23-27 h and 22-27h for α-and the β-isomers depending on the initial nominal concentration (Jonson and Toledon, 1993). Pesticides and other xenobiotics used in the shrimp farm and agriculture are posing a risk to a wide flora and fauna. The biological effects of endosulfan have been shown recently to be more significant than thought previously. Endosulfan poisoning was found to stimulate the central nervous system. Apart from its toxic effects on human health and animals, it is considered extremely toxic to beneficial microbial communities as reported by Buff et al., 1992.

A comparison of acute toxicity values indicated that endosulfan was more toxic to giant water prawn, Macrobranchium rosenbergii compared to Ametryne spp. as investigated by Lombardi et al., 2001. However the USEPA has classified endosulfan in category I as highly hazardous pesticide based on the results of another study in which the LD50 value was calculated as 18 to 160 mg/kg for rats, 7.36 mg/kg for mice and 77 mg/kg for dogs.The acute toxicity of α-isomer is three times more than of the β-isomer against mammals and the neurotoxicity caused by insecticide has been mainly attributed to α-isomer investigated and reported by Sutherland et al., 2004. Magesh and Kumaraguru 2006 reported on acute toxicity of endosulfan to the milk fish available at the south east coast of India.

Endosulfan degradation

Endosulfan is subjected to both biotic and abiotic degradation in the environment that may result in oxidation to the corresponding sulfate or hydrolysis in aquatic systems to endosulfan diol. In turn the diol may degrade further to endosulfan ether, endosulfan hydroxyether or endosulfan lactone (Walse et al., 2003).

MICROBIAL ENDOSULFAN DEGRADATION

In an international programme in 1984 on chemical safety held in WHO Geneva, reported on the environmental health criteria of endosulfan and according to it, it is broken down by photolysis, hydrolysis and biodegradation. Microbial degradation plays a major role in degradation of endosulfan contaminated sites in the environment(Shetty et al., 2000). Microbial degradation or microorganisms play an important role in the pesticide degradation in the natural environment. Microorganisms that are capable of degrading pesticides detoxify pesticides and influence the fate of pesticides in the environment.

In general, microorganisms demonstrate considerable capacity for the metabolism of many pesticides. Although they are capable of catalyzing similar metabolic reactions as mammals and plants, they possess the unique ability to completely mineralize many aliphatic, aromatic, and heterocyclic compounds. There are two major types of microbial degradation of organic chemicals. The first, termed catabolism, is a type of degradation in which the organic chemical or a portion there of is completely degraded (e.g. mineralized) and the energy or nutrient gained contributes to cell growth. The second, incidental metabolism or co-metabolism, involves the partial degradation of an organic chemical with no net benefit to the organism, the compound being merely caught up in some metabolic pathway during the normal metabolic activities of the microorganisms (Racke, 1993).

Microbial communities define the magnitude and pathways of organic matter, nutrient and energy dynamics in aquatic systems. Lynch and Hobbie (1988) reported on the metabolic activities of indigenous microorganisms in the environment which cause extensive modifications in the structure and toxicological properties of the contaminants resulting in the complete conversion of xenobiotics into innocuous inorganic end products.

Martens (1976), reviewed on 16 fungi, 15 bacteria, and 3 actinomycetes which were capable of metabolizing more than 30% of the applied endosulfan. The major metabolites detected were endosulfate, formed by oxidation of the sulfite group, and endodiol, formed by hydrolysis of the ester bond. The majority of highly active fungi formed endosulfate as the major metabolite, whereas the majority of active bacteria formed endodiol. Kumar and Philip, 2006 isolated three novel bacterial species namely, Staphylococcus sp., Bacillus circulans – I, and Bacillus circulans – II, from contaminated soil of a pesticide manufacturing industry and assessed their potential for the degradation of aqueous endosulfan in aerobic and facultative anaerobic conditions. After four weeks of incubation using mineral salt medium, mixed bacterial culture was able to degrade 71.82 ± 0.2% and 76.04 ± 0.2% of endosulfan in aerobic and facultative anaerobic conditions respectively, with an initial endosulfan concentration of 50mgl-1. Among the three species, Staphylococcus sp. utilized more beta endosulfan compared to alpha endosulfan in facultative anaerobic system, whereas Bacillus circulans – I and Bacillus circulans – II utilized more alpha endosulfan compared to beta endosulfan in aerobic system.

Jayashree and Vasudevan (2007), investigated on the effect of Tween 80 added to the soil on the degradation of endosulfan by a Pseudomonas aeruginosa strain at different pH of about 7.0 and 8.5. The addition of synthetic surfactant Tween 80 enhanced the solubility and degradation of endosulfan. A significant degradation (94%) was observed in pH 8.5 in the Tween 80 added soil. The degradation of both the isomers were observed and accompanied with the formation of endodiol and endosulfan sulfate.

Kwon et al.(2005) , isolated a bacterium, Klebsiella oxytoca capable of degrading endosulfan sulfate as well as endosulfan using mineral salt medium containing 150 mg l-1 of endosulfan or 173 mgl-1 of endosulfan sulfate. The biomass was rapidly increased to an optical density of 1.9 at 550 nm in 4 days and the degradation constants for α- and βendosulfan, and endosulfan sulfate were respectively 0.3084, 0.2983 and 0..2465 per day. Balerao and Puranik (2007) investigated on the soil fungi in endosulfan degradation. Totally 16 fungal isolates were obtained by enrichment of soil samples that had been exposed to endosulfan earlier. On the basis of tolerance to endosulfan, Aspergillus niger was selected as a potential organism which could tolerate 400 mg ml-1 of endosulfan. Complete disappearance of endosulfan was seen on 12 days of incubation. Evolution of carbon dioxide during endosulfan metabolism indicated the complete mineralization of endosulfan. Change in pH of culture broth to acidic range supported the biological transformation. Thin layer chromography (TLC) analyses revealed the formation of various intermediates of endosulfan metabolism including endosulfan diol, endosulfan sulfate and an unidentified metabolite.

Awasthi et al., 2000 examined a bacterial co-culture consisting of two strains of Bacillus sp., isolated from a contaminated site by selective enrichment technique using endosulfan as the sole carbon source which degraded endosulfan within 15 days in a culture flask, with a simultaneous increase in the bacterial biomass. Sutherland et al., 2002 isolated an endosulfan degrading bacterial strain designated as ESD from soil inoculums after repeated culture withthe insecticide as the sole source of sulfur. Enrichment of the sample was done for isolation of selective bacteria for effective degradation of endosulfan. Awasthi et al. (2003), studied the detoxification of α and β- isomers of endosulfan by a defined co-culture of two Bacillus strains. The degradation of both isomers was accompanied by the formation of endosulfan diol and endosulfan lactone. Accumulation of the endosulfan sulphate was not observed during the reaction with either of the isomers. Kwon et al., 2002 isolated bacteria that degraded endosulfan without formation of endosulfan sulfate from various soil samples using endosulfan as the sole carbon and energy source. Among the 40 isolated bacterial strains KE-1 identified as Klebsiella pneumonia showed superior endosulfan degrading activity. Analysis of culture pH, growth, free sulfate and endosulfan and its metabolites demonstrated that KE-1 biologically degraded 8.72μg of endosulfan per ml per day when it was incubated with 93.9 μg ml-1 of endosulfan for 10 days and the process was without the toxic formation of endosulfan sulfate.

Kumar et al., 2007 determined that two mixed bacterial cultures i.e., Stenotrophomonas maltophilia and Rhodococcus erythropolis were able to degrade respectively 73% and 81% of α and β endosulfan with endodiol intermediate, after two weeks of incubation. The toxicity studies of endosulfan before and after degradation were carried out using micronucleus assay on human polymorphonuclear cells, S. maltophilia showed better degradation efficiency compared to that of R. erythropolis. Li et al., 2009 isolated Achromobacter xylosoxidans CS5 capable of utilizing endosulfan as the sole carbon, sulfur and energy source from the activated sludge of Jiangsu, China.

Analysis of culture pH, cells growth and residual endosulfan demonstrated that CS5 could degrade more than 24.8 mg/l α endosulfan and 10.5 mg/l β endosulfan after 8 days in aqueous medium, with the formation of endosulfan diol and endosulfan ether as the major metabolites. Cell-free extract of strain CS5 was able to metabolize endosulfan rapidly, and the degradative enzymes were constitutively expressed. Inoculation of strain CS5 was found to promote the removal of endosulfan in soil and also suggested that A. xylosoxidans CS5 might degrade endosulfan by a non-oxidative pathway. To assess the degradation of endosulfan the chloride content was estimated according to Bergman and Sanik (1957) method. Kumar et al., (2008) analyzed the endosulfan degradation rate using GCMS for quantitative and qualitative analysis of the compound and its residues.

REFERENCES

1. Amaraneni, S. R. (2002): Persistence of pesticides in water, sediment and fish from fish farms in Kolleru Lake, India. J. Agric. Food Chem., 82: 918 - 923.
2. Arnold, S. F., Klotz, D. M., Collins, B. M., Vonier, P. M., Guillette, L. J. J. and McLachlan, J.A. (1996): Synergistic activation of estrogen receptors with combination of environmental chemicals. Sci., 272:1489-1492.
3. Awasthi, N., Ahuja, R. and Kumar, A. (2000): Factors influencing the degradation of soilapplied endosulfan isomers. Soil Biol. Biochem., 32: 1697-1705.
4. Awasthi, N., Singh, A. K., Jain, R. K. and Khangarot, B. S. (2003): Degradation and detoxification of endosulfan isomers by a defined co-culture of two Bacillus strains. Appl. Microbial. Biotechnol., 62: 279 - 283.
5. Bergmam, J. G. and Sanik, J. (1957): Determination of trace amounts of chlorine in naptha. Anal. Chem., 29:241-243. Universal Journal of Environmental Research and Technology 375 Vivekanandhan et al.
6. Bhalerao, T. S. and Puranik, P. R. (2007): Biodegradation of organochlorine pesticide, endosulfan, by a fungal isolate, Aspergillus niger. Inter. Biodet. Biodeg., 59(4):315-321.

7. Buff, K.D., Mano, M. S. and Langenbach, T. (1992): Effect of endosulfan on Azospirillum lipoferum growth, morphology, nitrogenase activity and protein binding. Appl. Environ. Microbiol., 58(9):3173-3176.

8. Chaudhuri, K., Selvaraj, S. and Pal, A. K. (1999): Studies on the genotoxicology of endosulfan in bacterial systems. Mutat. Res. Genet. Toxicol. Environ. Mut., 439: 63-67.

9. Daly, H., Doyen, J. T. and Purcell, A. H. (1998): Introduction to insect biology and diversity, 2nd edn. Oxford University Press, New York. 14: 279300.

10. Delorenzo, M. E., Taylor, L. A., Lund, S. A., Pennington P. L., Strozier, E. D. and Fulton, M. H. (2002): Toxicity and bioconcentration potential of the agricultural pesticide endosulfan in phytoplankton and zooplankton. Arch. Environ. Contam. Toxicol., 42:173-181.

11. Goebel, H., Gorbach. S., Knauk, W., Rimpau, R. H. and Huttenbach, H. (1982): Properties, effect, residues and analytics of the insecticide endosulfan. Residue. Rev., 83:1-22.

12. Guerin, T. F. (1999): The anaerobic degradation of endosulfan by indigenous microorganisms from low-oxygen soils and sediments. Environ. Poll., 06:13-21.

13. Guerin, T. F. (2001): Abiological loss of endosulfan and related chlorinated organic compounds from aqueous systems in the presence and absence of oxygen. Environ. Poll., 15:219-230.

14. Guerin, T. F. and Kennedy, I. R. (1992): Distribution and dissipation of endosulfan and related cyclodienes in sterile aqueous systems: implications for studies on biodegradation. J. Agri. Food Chem., 40:2315-2323.

15. Hatzios, K. K. (1991): Biotransformation of herbicides in higher plants, in: R. Grover, A. J. Cessna (Eds.). Environ. Chem. Herb., 2:141-185.

16. Jayashree, R and Vasudevan, N. (2007): Effect of tween 80 added to the soil on the degradation of endosulfan by Pseudomonas aeroginosa. Int. J. Environ. Sci. Tech., 4(2): 203-210.

17. Jonson, C. M. and Toledo, M. C. F. (1993): Bioaccumulation and elimination of endosulfan in the fish yellow tetra

(HyphessobryconBifasciatus). Bull. Environ. Contam. Toxicol., 50:572-577.

18. Kaur, I., Matur, R. P., Tandon, S. N. and Dureja, P. (1998): Persistence of endosulfan (technical) in water and soil. Environ. Technol., 19: 115-119.
19. Keith, L. H. and Telliard, W. A. (1979): Priority pollutants1: A perspective view. Environ. Sci. Technol., 13: 416-423.
20. Kumar, K., Devi, S. S., Krishnamurthi, K., Kanade, G. S. and Chakrabarti, T. (2007): Environment and isolation of endosulfan degrading and detoxifying bacteria. Chemo., 68(3):479-488.
21. Kumar, M. and Philip, L. (2006): Enrichment and isolation of a mixed bacterial culture for complete mineralization of endosulfan. J. Environ. Sci. Health, 41: 81-96.
22. Kumar, M., Lakshmi, C. V. and Khanna, S. (2008): Biodegradation and bioremediation of endosulfan contaminated soil. Biores. Technol., 99: 3116-3122.
23. Kwon, G. S., Kim, J. E., Kim, T. K., Sohn, H. Y., Koh, S. C., Shin, K. S. and Kim, D. G. (2002): Klebsiella pneumonia KE-1 degrades endosulfan without formation of the toxic metabolite, endosulfan sulfate. FEMS Microbiol. Lett., 215:255-259.
24. Kwon, G. S., Sohn, H. Y., Shin, K. S., Kim, E. and Seo, B. I. (2005): Biodegradation of the organochlorine insecticide, endosulfan and the toxic metabolite, endosulfan sulfate, by Klebsiella oxytoca KE-8. Appl. Microbiol. Biotechnol., 67:845-850.
25. Li, W., Dai, Y., Xue, B., Li, Y., Peng, X., Zhang, J. and Yan, Y. (2009): Biodegradation and detoxification of endoulfan in aqueous medium and soil by Achromobacter xylosoxidans strain CS5. Jour. Hazar. Mat., 167(1-3):209-216.
26. Lobe, J. (2006): "WHO urges DDT for malaria control Strategies," Inter Press Service, cited from Commondreams.org.
27. Lombardi, J. V., Neto, G. J. M., Garcia, A. L. B., Marques, H. L. A. and Kubo, E. (2001): Acute toxicity of the pesticides endosulfan and ametryne to the fresh water prawn Macrobranchim rosenbergii De man. Bull. Environ. Contam. Toxicol., 67:665-671.
28. Lynch, J. M. and Hobbie, J. E. (1988): Microorganisms in action:

Concepts and applications in microbial ecology. Bla. Sci. Pub., 322-347.

29. Magesh, S. and Kumaraguru, A. K. (2006): Acute toxicity of endosulfan to the milkfish, Chanos chanos, of the southeast coast of India. Bull. Environ. Contam. Toxicol., 76:622-628. Universal Journal of Environmental Research and Technology 76 Vivekanandhan et al.
30. Mansingh, A. and Wilson, A. (1995): Insecticide contamination of Jamaican environment. Baseline studies on the status of insecticidal pollution of Kingston Harbour. Mar. Poll. Bull., 30:640-645.
31. Martens, R. (1976): Degradation of [8, 9-14C] Endosulfan by soil microorganisms. Appl. Environ. Microbial., 6:853-858.
32. Miles, C. J. and Pfeuffer, R. J. (1997): Pesticides in canals of South Florida. Arch. Environ. Contam. Toxicol., 32: 337-345.
33. Miller, G. T., (2002): Living in the Environment (12th Ed.). Belmont: Wadsworth/Thomson Learning.
34. Miller, G. T., (2004): Sustaining the Earth, 6th edition. Thompson learning, Inc. Pacific Grove, California. 9:211-216.
35. Muller, W. P. and Korte, F. (1975): Microbial degradation of benzo-(a)-pyrene, monolinuron and dieldrin in waste composting. Chem., 4:195198.
36. Paul, V. and Balasubramaniam, E. (1997): Effect of single and repeated administration of endosulfan on behavior and its interaction with centrally acting drugs in experimental animals: A mini review. Environ. Toxicol. Pharmacol., 3:151157.
37. Racke, K. D. (1993): Environmental fate of chlorpyrifos. Rev. Environ. Contam. Toxicol., 131:1-151.
38. Ramnarine and Kristy, (2008): Harmful elements in Sahara dust. Trinidad & Tobago express.
39. Rao, D. M. R. and Murty, A. S. (1980): Persistence of endosulfan in soils. J. Agri. Food. Chem., 28: 1099-1101.
40. Saiyed, H., Dewan, A., Bhatnaga, V., Shenoy, U., Shenoy, R. and Rajmohan, H. (2003): Effect of endosulfan on male reproductive development. Environ. Heal. Pers., 111: 1958-1962.

41. Sethunathan, N., Megharaj, M., Chen, Z. L., Lewis, B. D. W. and Naidu, R. (2002): Algal degradation of a known endocrine disrupting insecticide, α- endosulfan, and its metabolite, endosulfan sulfate, in liquid medium and soil. J. Agri. Food Chem., 52: 3030-3035.

42. Shetty, P. K., Mitra, J., Murthy, N. B. K., Namitha, K. K., Sovitha, K. N. and Raghu, K. (2000): Biodegradation of cyclodiene insecticide endosulfan by Mucor thermo hyalospora MTCC 1384. Curr. Sci., 79: 1381-1383.

43. Shimabukuro, R. H., (1985): Detoxification of herbicides, In: S.O. Duke (Ed.), We. Physio., 2: 215-240.

44. Siddique, T., Okeke, B. C., Arshad, M. and Frankenberger, W. T. Jr. (2003): Biodegradation kinetics of endosulfan by Fusarium ventricosum and a Pandoraea species. J. Agri. Food Chem., 51: 8015-8019.

45. Sinha, N., Narayan, R. and Saxena, D.K. (1997): Effect of endosulfan on testis of growing rats. Bull. Environ. Contam. Toxicol., 58: 79-86.

46. Sunderam, R. I. M., Cheng, M. H. and Thompson, G. B. (1992): Toxicity of endosulfan to native and introduced fish in Australia. Environ. Toxicol. Chem., 11: 1469-1476.

47. Sutherland, T. D., Home, I., Weir, K. M., Russel, R. J. and Oakeshott, J. G. (2004): Toxicity and residues of endosulfan isomers. Rev. Environ. Contam. Toxicol., 183: 99-113.

48. Sutherland, T., Horne, I., Lacey, M., Harcourt, R., Russell, R. and Oakeshott, J. (2002): Isolation and characterization of a Mycobacterium strain that metabolises the insecticide endosulfan. J. Appl. Microbiol., 93: 380-389.

49. Thurman, E. M., Goolsby, D. A., Meyer, M. T., Mills, M. S., Pomes, M. L. and Kolpin, D.W. (1992): A reconnaissance study of herbicides and their metabolites in surface water of the Midwestern United States using immunoassay and gas chromatography/mass spectrometry. Environ. Sci. Tech., 26(12): 2440-2447.

50. Walse, S. S., Scott, G. I. and Ferry, J. L. (2003): Stereoselective degradation of aqueous endosulfan in modular estuarine mesocosms: formation of endosulfan – hydroxycarboxylate. J. Environ. Moni., 5: 373-379.

51. Wan, M. T., Kuo, J. N., Buday, C., Schroeder, G., Aggelen, G. V. and Pasternak, J. (2005): Toxicity of α- and β- endosulfan and their formulated and degradation products to Daphnia manga, Hyalella azteca, Oncorhynchus mykiss, Oncorhynchus kisutch and biological implications in streams. Environ. Toxicol. Chem., 24: 1146-1154.

9
CHAPTER

ECOLOGICAL EFFECTS OF PESTICIDES

INTRODUCTION

Definition of ecology

At a community level, ecology can be defined as complex interactions that exist among interdependent organisms that cohabitate the same geographical area and with their environment (Johnson and Strinchcombe, 2007). At individual level, it entails the relationships that exist between that particular individual with numerous physical and biological factors. The physical environment along with organisms (biota) inhabiting a particular space make up an ecosystem. Some typical examples of ecosystems include a farm pond, a mountain meadow and rain forest. In a natural environment, an ecosystem follows a certain sequence of processes and events through the days, seasons and years. The processes include not only the birth, growth, reproduction and death of biota in that particular ecosystem, but also the interactions between species and physical characteristics of the geological environment. From these processes and interactions, the ecosystem gains a recognizable structure and function, and matter and energy are exchanged and cycled through the ecosystem. Over time, better adapted species come to dominate; entirely new species may change, perhaps in a new or altered ecosystem.

Organisation of the ecosystems

The basic level of ecological organisation is with the individual such as a single plant, insect or bird. The definition of ecology is based on the interactions of organisms with their environment. In the case of an individual, it would entail the relationships between that individual and numerous physical factors such as rain, sun, wind, temperature, nutrients. For biological such as other plants, insects, diseases, animals, etc. Factors. The next level of organization is the population that are no more than a collection of individuals of the same species within an area or region. Populations will mean the populations of humans, birch trees, or sunfish in a pond. Population ecology is concerned with the interaction of the individuals with each other and with their environment. The next, more complex, level of organization is the community. Communities are made up of different populations of interacting plants, animals, and microorganisms also within some defined geographic area. Different populations within a community interact more among themselves than with populations of the same species in other communities, therefore, there are often genetic differences between members of two different communities. The populations in a community have evolved together, so that members of that community provide resources such as food or nutrition and shelter for each other.

PESTICIDES IN THE MODERN WORLD - RISKS AND BENEFITS

The next level of organization is the ecosystem. An ecosystem consists of different communities of organisms associated within a physically defined space. For example, a forest ecosystem consists of animal and plant communities in the soil, forest floor, and forest canopy, along the stream bank and bottom, and in the stream. A stream bottom community, for example, will have various fungi and bacteria living on dead leaves and animal wastes, protozoans and microscopic invertebrates feeding on these microbes, and larger invertebrates such as worms and crayfish and for vertebrate's example are the turtles and catfish. Each community functions somewhat separately, but are also linked to others by the forest, rainfall, and other interactions. For example, the stream community is heavily dependent upon leaves produced in the surrounding trees falling into the stream, feeding the microbes and other invertebrates. Terrestrial ecosystems can be grouped into units of similar nature, termed biomes such as a "deciduous forest," "grassland," "coniferous forest," and or into a geographic unit, termed landscapes, containing several different types of ecosystems. Aquatic ecosystems are commonly categorized on the basis of whether the water is moving in either streams and river basins or still waters such as ponds, lakes, large lakes and whether the water is fresh, salty such asoceans, or brackish such as estuaries. Landscapes and biomes example are large lakes, river basins, and oceans which are subject to global threats of pollution of containing acid deposition, stratospheric ozone depletion, air pollution, the greenhouse effect and human activities such as soil erosion, deforestation, pesticides use.

Pesticides environmental contaminants

Pesticides released into the environment may have several adverse ecological effects ranging from long-term effects to short-lived changes in the normal functioning of an ecosystem. Despite the good results of using pesticides in agriculture and public health, their use is usually accompanied with deleterious environmental and public health effects. Pesticides hold a unique position among environmental contaminants due to their high biological toxicity which lead to acute and chronic diseases. Pesticides by definition are toxic chemical agents. A pesticide is usually capable of harming all forms of life other than the targeted pest species. On account of this behavior then, they can best be described as biocides are the one capable of killing all forms of life. Although some pesticides are described to be selective in their mode of action, their range of selectivity is only limited to the test animals.

Pesticides distribution in the environment

The term chemodynamics of pesticides refers to the study of the movement and transformation of pesticides as well as their fate in various compartments of the environment. The environment can be divided into four major compartments, namely; air, water, soil and biota (Fig. 9. 1). The widespread use and disposal of pesticides by farmers, institutions and the general public provide many possible sources of pesticides in the environment. Pesticides once released into the environment may have many different fates. Pesticides that are sprayed can move through the air and may eventually end up in other parts of the environment, such as in soil or water. Pesticides that are applied directly to the soil may be washed off the soil into nearby bodies of surface water or may percolate through the soil to lower soil layers and groundwater (Harrison, 1990). This incomplete list of possibilities suggests that the movement of pesticides in the environment is very complex with transfers occurring continually among different environmental compartments. In some cases, these exchanges occur not only between areas that are close together (such as a local pond receiving some of the herbicides applied on adjacent land) but also may involve transportation of pesticides over long distances. The worldwide distribution of DDT and the presence of pesticides in bodies of water such as the Great Lakes far from their primary use areas are good examples of the vast potential of such movement.

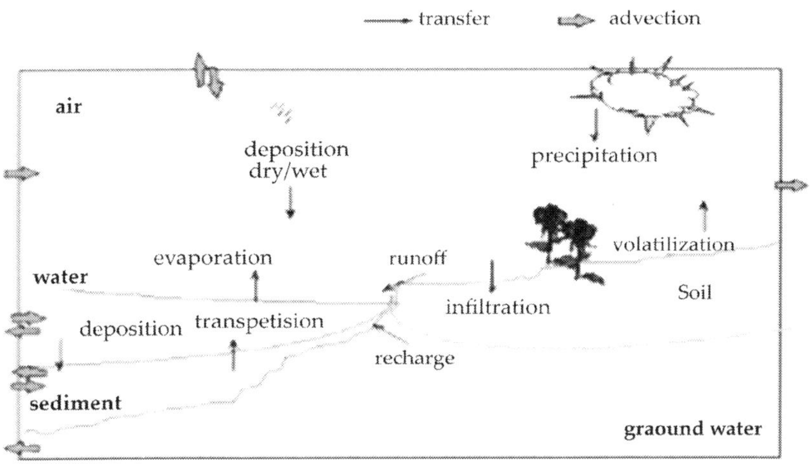

Figure 9.1: Distribution of pesticides in different environmental compartments

The major ecological effects of pesticides are described based on contemporary accumulated knowledge on the behavior of pesticides and the damage they cause to the ecosystem and the environment at large as a result of excessive use and/ or injudicious use of pesticides. The effects may range from minor deviation on the normal functioning of the ecosystem to the loss of species diversity in the ecosystem. Since organisms in the ecosystem live in a complex interdependent association with each other, the loss one key species may result in the collapse of the particular ecosystem. These effects are an important reason for the current strict regulations on the judicial use of pesticides.

The primary objective of using pesticides in the fields and the environment in general is to achieve a control of crop pests and disease vectors. This has been a deliberate human effort in a search for increasing agricultural yields and improving public health (Helweg, 2003). Pesticides applied to the environment have shown to have long term residual effects while others have shown to have acute fatal effects when not properly handled. Organochlorine pesticides for example have shown to be persistent in the environment, the result of which find their way to contaminate ground water, surface water, food products, air, soil and may affect human being through direct contact. Pesticides exposure to humans have been well documented to be the route cause of some diseases such as cancer, respiratory diseases, skin diseases, endocrine disruption, and reproduction disorders. It is this aspect of pesticide in the environment that has raised concern among environmental scientists to study their behaviour in the environment and then come out with a sound alternative so as to rescue the human population from their adverse effects. Fifty years (half a century) after Rachel Carson's eloquent warning to the world about the devastating effect pesticides have on birds and beneficial insects, pesticides continue to be a pervasive and insidious threat to the world's ecosystems. A massive chemical assult on our environment is launched each year. This poisonous barrage aggravates other pressures on our ecosystems such as expanding suburbarn development and dammed rivers, threatening the survival of many birds, fish, insects, and small aquatic organisms that form the basis of the food web. More generally, pesticides reduce species diversity in the animal kingdom and contribute to population decline in animals and plants by destroying habitats, reducing food supplies and impairing reproduction (Kegley, et al, 1999).

Loss of species diversity

Organisms in ecosystems exist in complex interdependent associations such that losses of one keystone species as a result of pesticides that can cause a far reaching and unpredictable effects. A keystone species is a species that is disproportionately connected to more species in the food-web. The many connections that a keystone species holds means that it maintains the organization and structure of entire communities. The loss of a keystone species results in a range of dramatic cascading effects that alters trophic dynamics, other food-web connections and can cause the extinction of other species in the community. Sea otters (*Enhydra lutris*) for example, are known to be keystone species in marine ecosystems that limits the density of sea urchins (Mills, et al, 1993). A pesticide may eliminate a species essential to the functioning of the entire community, or it may promote the dominance of undesired species or it may simply decrease the number and variety of species present in the community. This may disrupt the dynamics of the food webs in the community by breaking the existing dietary linkages between species. The literature on pest control lists many examples of new pest species that have developed when their natural enemies are killed by pesticides. This has created a further dependence on pesticides. Finally, the effects of pesticides on the biodiversity of plants and animals in agricultural landscapes, whether caused directly or indirectly by pesticides, constitute a major adverse environmental impact of pesticides.

Effects involving pollinators

Some natural pollinators, such as honeybees and butterflies, are very sensitive to pesticides. Pesticides can kill bees and are strongly implicated in pollinator decline, the loss of species that pollinate plants, including through the mechanism of Colony Collapse Disorder (Hackenberg, 2007), in which worker bees from a beehive or Western honey bee colony abruptly disappear. Application of pesticides to crops that are in bloom can kill honeybees, which act as pollinators. The USDA and USFWS estimate that US farmers lose at least $200 million a year from reduced crop pollination because pesticides applied to fields eliminate about a fifth of honeybee colonies in the US and harm an additional 15% (Miller, 2004). Since these are important pollinators of both crops and native plants, reduced number of natural pollinators can therefore result into reduced seed and fruit production. This is both an ecological effect as well as economical effect. Bees are extremely important in the pollination of crops and wild plants, and although pesticides are screened for toxicity to bees, and the use of pesticides toxic to bees is permitted only under stringent conditions, many bees are killed by pesticides, resulting in the considerably reduced yield of crops dependent on bee pollination.

Figure 9.2: Butterflies and bees are the representative natural pollinating agents for plants.

EFFECTS ON NUTRIENT CYCLING IN ECOSYSTEM

A large proportional of pesticides used in the environment ultimately reaches the soil where soil building processes and the cycling of nutrients back into living plants is accomplished. Pesticides can affect the soil organisms involved in these processes directly or indirectly hence interfering with the natural nutrient cycling in the ecosystem.

Nitrogen fixation, which is required for the growth of higher plants, is hindered by pesticides in soil. The insecticides DDT, methyl parathion, and especially pentachlorophenol have been shown to interfere with legume-rhizobium chemical signaling. Reduction of this symbiotic chemical signaling results in reduced nitrogen fixation and thus reduced crop yields (Rockets, 2007). Root nodule formation in these plants saves the world economy $10 billion in synthetic nitrogen fertilizer every year (Fox, 2007). When the natural nutrient cycling (figure 10.3) in the ecosystem is interfered in any way by pesticides or other sources of pollution, it will lead to decline in soil fertility and soil productivity.

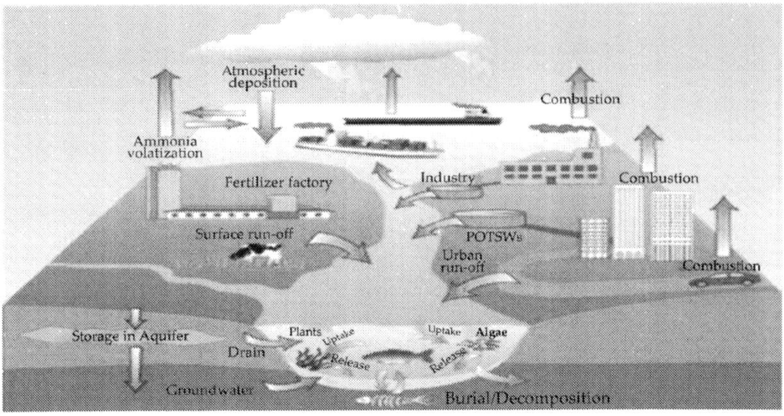

Figure 9.3: Nutrient cycling in ecosystem

EFFECTS ON SOIL EROSION, STRUCTURE AND FERTILITY

Many of the chemicals used in pesticides are persistent soil contaminants, whose impact may endure for decades and adversely affect soil conservation. The use of pesticides decreases the general biodiversity in the soil. Not using the chemicals results in higher soil quality (Johnson, 1986), with the additional effect that more organic matter in the soil allows for higher water retention. This helps increase yields for farms in drought years, when organic farms have had yields 20-40% higher than their conventional counterparts. A smaller content of organic matter in the soil increases the amount of pesticide that will leave the area of application, because organic matter binds to and helps break down pesticides (Lotter, et al, 2003). Herbicides for example can reduce vegetative cover of the ground, thus promoting soil erosion via runoff and wind. Soil erosion deforms the soil structure and therefore creates an imbalance in soil fertility. A bare land with poor soil structure and poor soil fertility cannot support the growth of plants on it. Ecologically this land cannot support other forms of life in it hence may lead to the collapse of the particular ecosystem.

EFFECTS ON WATER QUALITY

Pesticides applied in the environment can find their way into water bodies either from the air or by runoff or by percolation to groundwater. There are four major routes through which pesticides can reach the water bodies: it may drift outside of the intended area when it is sprayed, it may percolate, or leach, through the soil, it may be carried to the water as runoff, or it may be spilled, for example accidentally or through negligence. They may also be carried to water by eroding soil. Factors that affect a pesticide's ability to contaminate water include its water solubility, the distance from an application site to a water body, weather, soil type, presence of a growing crop, and the method used to apply the chemical. Once pesticides enter water bodies they have a potential to cause harmful effects on human health, aquatic organisms and can cause disruptions of the aquatic ecosystems. This may result into a loss in fish production in streams and large water bodies especially where fishing is one among the major economic activities of a particular community. In the United States for example, pesticides were found to pollute every stream and over 90% of wells sampled in a study by the US Geological Survey (Gillion, et al, 2007). Pesticide residues have also been found in rain and groundwater. Studies by the UK government showed that pesticide concentrations exceeded those allowable for drinking water in some samples of river water and groundwater (Bingham, 2007).

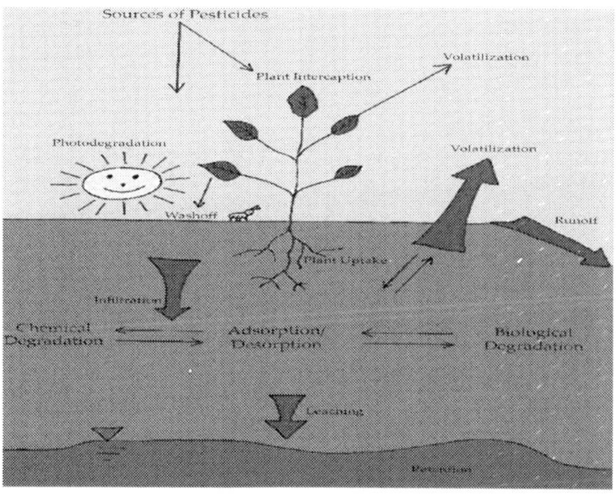

Figure 9.4: Pesticieds pathways in contaminating water bodies (Heather, et al, 1997)

EFFECTS ON BIRDS

Pesticides have had some of their most striking effects on birds, particularly those in the higher trophic levels of food chains, such as bald eagles, hawks, and owls. These birds are often rare, endangered, and susceptible to pesticide residues such as those occurring from the bioconcentration of organochlorine insecticides through terrestrial food chains. Pesticides may kill grain- and plant-feeding birds, and the elimination of many rare species of ducks and geese has been reported. Populations of insect-eating birds such as partridges, grouse, and pheasants have decreased due to the loss of their insect food in agricultural fields through the use of insecticides. The loss of even a few individuals from rare, endangered or threatened species pushes the entire species close to extinction. Some pertinent examples associated with birds' kills as a result of pesticides include the insecticides diazinon and carbofuran which are well document as causing bird kills in many parts of the world (Kegley et al, 1999). Organochlorine insecticides such as DDT are also well documented to have continued impairing avian reproduction even after years of banned use.

Figure 9.5: A group of birds that died as a result of pesticides application.

Most bird kills go undocumented, with reported kills representing only a small fraction of actual bird mortality due to pesticides. Birds exposed to sublethal doses of pesticides are also afflicted with chronic symptoms that affect their behaviour, reproduction, and nervous system. Weight loss, increased susceptibility to predation, decreased disease resistance, lack of interest in mating and defending territory, and abandoning of nestlings have been observed as side effects of pesticides exposure.

EFFECTS ON FISH AND OTHER AQUATIC ORGANISMS

A major environmental impact has been the widespread mortality of fish and marine invertebrates due to the contamination of aquatic systems by pesticides. This has resulted from the agricultural contamination of waterways through fallout, drainage, or runoff erosion, and from the discharge of industrial effluents containing pesticides into waterways. Historically, most of the fish in Europe's Rhine River were killed by the discharge of pesticides, and at one time fish populations in the Great Lakes in USA became very low due to pesticide contamination. Additionally, many of the organisms that provide food for fish are extremely susceptible to pesticides, so the indirect effects of pesticides on the fish food supply may have an even greater effect on fish populations. Some pesticides, such as pyrethroid insecticides, are extremely toxic to most aquatic organisms. It is evident that pesticides cause major losses in global fish production. Furthermore, recent laboratory studies of endosulfan and fenitrothion in the tilapia species from Lake Victoria in Tanzania indicated a high capacity of the species to absorb the two pesticides from water with rapid distribution in the organs each with a bioaccumulation factor of 33 and 346 L/ kg fresh weight respectively (Henry, 2003). Multiple pesticides contamination are very common in water and sediments, frequently at concentrations exceeding the lethal limits for many species of zooplankton, small species of animals eaten by fish. Because of the significant high water solubility of the insecticides diazinon and chlorpyrifos and the herbicides simazine, diron, and EPTC are found most commonly in water bodies and have been associated with fish kills and decline in zooplankton population in aquatic environment.

Figure 9.6: Spraying an aquatic herbicide

EFFECTS ON FROGS AND OTHER AQUATIC AMPHIBIANS

Atrazine being one of the world`s most used pesticide has recently reported by laboratory studies to have a effect on changing male frogs (African clawed frog; Xenopus laevis). Adult frogs exposed to atrazine turn female one in ten (10%). These male frogs are missing testesterone and all things controlled by testesterone including sperm production. So their fertility is as low as 10 percent when treated in isolation, but when treated with normal males, they stand a zero chance of reproducing. Although 10 percent of these mutant females can successful mate with male frogs, their offspring are all male because they are genetically male frogs. The ultmate effect of this is that the sex ratios of frogs is badly skewed and this is very dangerous for the survival of that species (Hayes et al, 2010). Kihansi spray toads is one among the world`s rarest amphibian species that was close to extinction from their natural environment in Tanzania. The species was first discovered in 1996 during an environment impact study for a large new hydroelectic dam in Udzungwa Mountains in Southern Tanzania. The toads lived exclusively in a five acre zone under spray of a waterfall from Udzungwa Mountains, hence the name Kihansi spray toads. Among other reasons that contributed to the decline is the use of pesticides in the environment. To rescue this rare species of toads, a colony of them was taken to Bronx zoo and Toledo zoo in USA where they were reared and bread in laboratories for 10 years.

Figure 9.7: Kihansi spray toads from Kihansi Gorge in Tanzania

Pesticides disrupt the natural balance between pest and predator insects Broad spectrum pesticides such as organochlorine, organophosphorus and carbamate insecticides destroy both pest and beneficial organisms indiscriminately, thus upsetting the natural balance between pests and predator insects. Beneficial organisms serve many valuable functions in an agricultural ecosystem including pollination, soil aeration, nutrient cycling, and natural pest control through pest-predator relationship. Application of insecticides indiscriminately kills both pests and beneficial organisms. Pest populations often recover rapidly because of their lager numbers and ability to develop resistance, but beneficial organisms do not, resulting in a resurgence of the target pest as well as secondary pests that reproduce rapidly without natural predator to check down their numbers. This prompts an escalation in the use of more pesticides by the farmers in an attempt to control them and boost their harvest.

Figure 9.8: Aerial spraying of pesticides onto the crops using an aircraft

Pesticides cause pest rebound and secondary pest outbreaks

Non-target organisms, organisms that the pesticides are not intended to be killed, can be severely affected by the use of pesticides. In some cases, where a pest insect has some controls from a beneficial predator or parasite, an insecticide application can kill both pest and beneficial populations. A study comparing biological pest control and use of pyrethroid insecticide for diamondback moths, a major cabbage family insect pest, showed that, the insecticide application created a rebounded pest population due to loss of insect predators, whereas the biological control did not show the same effect (Muckenfuss, et al 1990). Likewise, pesticides sprayed in an effort to control adult mosquitoes, may temporarily depress mosquito populations, however they may result in a larger population in the long run by damaging the natural controlling factors. This phenomenon, wherein the population of a pest species rebounds to equal or greater numbers than it had before pesticide use, is called pest resurgence and can be linked to elimination of predators and other natural enemies of the pest (Daly, et at, 1998) The loss of predator species can also lead to a related phenomenon called secondary pest outbreaks, an increase in problems from species which were not originally very damaging pests due to loss of their predators or parasites (Daly, et at, 1998). An estimated one-third of the 300 most damaging insects in the US were originally secondary pests and only became a major problem after the use of pesticides (Miller, 2004). In both pest resurgence and secondary pest outbreaks, the natural enemies have been found to be more susceptible to the pesticides than the pests themselves, in some cases causing the pest population to be higher than it was before the use of pesticide.

EFFECTS ON HUMAN BEINGS

Pesticides can enter the human body through inhalation of aerosols, dust and vapor that contain pesticides; through oral exposure by consuming contaminated food and water; and through dermal exposure by direct contact of pesticides with skin (Sacramento, 2008). Pesticides are sprayed onto food, especially fruits and vegetables, they secrete into soils and groundwater which can end up in drinking water, and pesticide spray can drift and pollute the air. The effects of pesticides on human health are more harmful based on the toxicity of the chemical and the length and magnitude of exposure (Lorenz, 2009). Farm workers and their families experience the greatest exposure to agricultural pesticides through direct contact with the chemicals. But every human contains a percentage of pesticides found in fat samples in their body. Children are most susceptible and sensitive to pesticides due to their small size and underdevelopment. The chemicals can bioaccumulate in the body over time. Exposure to pesticides can range from mild skin irritation to birth defects, tumors, genetic changes, blood and nerve disorders, endocrine disruption, and even coma or death (Miller, 2004).

Pests may evolve to become resistant to pesticides as a result of continued use of pesticides in a particular environment. Many pests will initially be very susceptible to pesticides, but some with slight variations in their genetic makeup they become resistant and therefore survive to reproduce. Through natural selection, the pests may eventually become very resistant to the pesticide. Pest resistance to a pesticide is commonly managed through pesticide rotation, which involves alternating among pesticide classes with different modes of action to delay the onset of or mitigate existing pest resistance. Tank mixing pesticides is the combination of two or more pesticides with different modes of action in order to improve individual pesticide application results and delay the onset of or mitigate existing pest resistance.

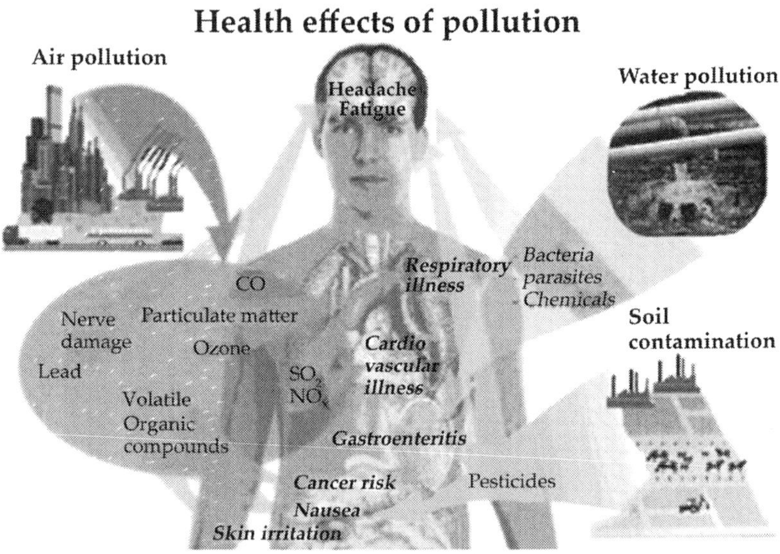

Figure 9.9: Impacts of pesticides on human health

Pesticides destroy the delicate balance between species that characterize the functioning ecosystem. With pesticides now being found routinely in drinking water, on food and in the air, we are all taking part in an experiment in pesticide exposure on a global scale, but without the benefit of an exposed control group for comparison. For that matter we are likely not be able to quantify the exact risk of these exposures. Because we cannot know for certain the consequences of the expanding pesticides use, the rational and most protective course of action is to take a precaution approach phasing out the use of the most dangerous pesticides, reducing our reliance on toxic chemicals for pest control and promoting ecologically based pest management. The adverse effects of pesticides on humans and wildlife have resulted in research into ways of reducing pesticide use. The most important of these is the concept of integrated pest management (IPM), first introduced in 1959. This combines minimal use of the least harmful pesticides, integrated with biological and cultural methods of minimizing pest losses. It is linked with using pesticides only when threshold levels of pest attacks have been identified. There is also a move toward sustainable agriculture which aims to minimize use of pesticides and fertilizers based on a systems approach.

There has been a growing concern recently on the promotion of organic farming which emphasize on techniques such as crop rotation, green manure, compost and biological methods of pest control to maintain soil productivity. Organic farming strictly excludes the use of manufactured fertilizers, pesticides, plant growth regulators, livestock antibiotics, food additives, and genetically modified organisms. Organic foods resulting from organic farming are deemed free from pesticides and hence providing an alternative source of quality and safe food in the future. By promoting the use of organic foods means will push the farmers to opt for organic farming. Market forces are a powerful incentive to encourage famers to go organic. Pesticides manufacturers should conduct long-term studies on ecosystem-wide impacts to demonstrate that a pesticide has no adverse effects before allowing it to be registered for use in the environment. The fact that present regulations view a pesticide as innocent until proved guilty is detrimental to the environment health. It is critical to know more about the long-term ecological effects of a pesticide before it is released to the environment. Using a combination of prior gained field experience with the existing pesticides and applying fundamental chemodynamic principles to newly developed compounds, we can now predict with some degree of accuracy the fate of new chemicals before they are even used in the environment.

REFERENCES

1. Johnson, M. T.; Strinchcombe, J. R. (2007). "An emerging synthesis between community ecology and evolutionary biology". Trends in Ecology and Evolution 22 (5): 250–7
2. Harrison, S. A. (1990). The Fate of Pesticides in the Environment, Agrochemical Fact Sheet # 8, Penn, USA Helweg, C. et al (2003). Fate of pesticides in surface waters, Laboratory and Field Experiments; Ministry of Environment, Danish Environmental Protection Agency,. Pesticides Research No. 68.
3. Kegley, S. et al (1999). Disrupting the Balance, Ecological Impacts of Pesticides in California, California, USA

4. Hackenberg D (2007). "Letter from David Hackenberg to American growers. Plattform Imkerinnen Austria.
5. Mills, L. S. Et al (1993) The Keystone-Species Concept in Ecology and Conservation, BioScience 43, (4), 219-224
6. Miller, G. T. (2004), Sustaining the Earth, 6th edition. Thompson Learning, Inc. Pacific Grove, California, USA
7. Rockets, R. (2007). Down On the Farm, Yields, Nutrients And Soil Quality Fox, J. E, et al (2007). "Pesticides reduce symbiotic efficiency of nitrogen-fixing rhizobia and host plants". Proceedings of the National Academy of Sciences, USA
8. Johnston, A. E. (1986). "Soil organic-matter, effects on soils and crops". Soil Use Management 2, 97–105
9. Lotter DW, Seidel R, and Liebhardt W (2003). "The performance of organic and conventional cropping systems in an extreme climate year". American Journal of Alternative Agriculture 18: 146–154
10. Gilliom, R.J. et al (2007). The Quality of our nation's waters: Pesticides in the nation's streams and ground water, 1992–2001. US Geological Survey
11. Bingham, S (2007), Pesticides in rivers and groundwater. Environment Agency, UK. Heather, B. et al (1997). Movement of Pesticides and Best Management Practices, Ground Water Pollution Primer, Virginia, USA
12. Henry, L. (2003) Levels of some Pesticides in Environmental Samples from Southern Lake Victoria and its Catchments and their Chemodynamics in Tilapia Species, Water and Sediments Under Experimental Conditions,
13. Tanzania Hayes et al, (2010) The cause of global amphibians decline: a developmental endocrinologist perspective, Journal of experimental biology, 213 (6) 921
14. Muckenfuss A. E. et al, (1990). Natural mortality of diamondback moth in coastal South Carolina Clemson University, Coastal Research and Education Center
15. Daly, H. et al (1998). Introduction to insect biology and diversity, 2nd edition, Oxford University Press, New York, USA

16. Sacramento, C. A. (2008). Department of Pesticide Regulation "What are the Potential Health Effects of Pesticides?" Community Guide to Recognizing and Reporting Pesticide Problems, 27-29
17. Lorenz, E. S. (2009) "Potential Health Effects of Pesticides." Ag Communications and Marketing. 1-8

10
CHAPTER

PESTICIDES AND THE LOSS OF BIODIVERSITY

WHAT IS BIODIVERSITY?

Charles Darwin and Alfred Wallace were among the first scientists to recognise the importance of biodiversity for ecosystems. They suggested that a diverse mixture of crop plants ought to be more productive than a monoculture (Darwin & Wallace 1858). Though there are exceptions, recent studies confirm the idea that an intact, diverse community generally performs better than one which has lost species (Chapinetal2002). Ecosystem stability (resilience to disturbance) seems to arise from groups of connected species being able to interact in more varied positive and complimentary ways (Tilman 2002).Biological diversity manifests itself at different levels. It includes the diversity of ecosystems, species, populations, and individuals. In an ecosystem, interdependent populations of various species deliver'services'such as the supply of food and soil resources, or the retention and cycling of nutrients, water and energy. Although it seems that the average species loss can affect the functioning of a wide variety of organisms and ecosystems, the magnitude of effect depends on which particular species is becoming extinct (Cardinaleetal2006). Communities of different animal and plant species perform vital functions within ecosystems. In general, communities which have higher diversity tend to be more stable.

Why conserve threatened species?

Rachel Carson provided clear evidence of the far-reaching environmental impact of pesticides in her pioneering work 50 years ago. In 'Silent spring' she showed that organochlorines, a large group of insecticides, accumulated in wildlife and the food chain. This had a devastating effect on many species. Only a decade after the 'green revolution' began it became obvious that large-scale spraying of pesticides was causing serious damage. In 1963, Rachel Carson emphasised human dependence on an intact environment: "But man is a part of nature, and his war against nature is inevitably a war against himself", (CBS 1963).Human well-being depends on the services delivered by intact ecosystems. While biodiversity loss is in itself a cause for concern, biodiversity conservation also aims to sustain humanity. People's livelihoods ultimately depend on biological resources. Thus lacking progress towards the target of the Convention on Biological Diversity, "to achieve by 2010 a significant reduction of the current rate of biodiversity loss" could undermine achievement of the Millenium Development Goals and poverty reduction in the long term (Sachset al2009). The 2010 target has inspired action but will not be fully attained. Biodiversity loss and degradation of ecosystems have increasingly dangerous consequences for people, and may threaten some societies' survival (IUCN 2010).

When EU cereal yield was doubled it resulted in the loss of half the plant species and one-third of carabid beetles and farmland bird species. Of the components of agricultural intensification, pesticide use, especially insecticides and fungicides, had the most consistently negative effects on species diversity, and insecticides also reduced the potential for biological pest control (Geigeretal 2010). In the EU, up to 80% of protected habitat types and 50% of species of conservation interest now have an unfavourable conservation status.Much greater effort is needed to reverse the decline in threatened species or habitats on a larger scale (EC 2008).A'business-as-usual'scenario would mean that the current decline of biodiversity will continue and even accelerate, and by 2050 a further 11% of natural areas which existed in 2000 will be lost, while

40% of land currently under low-impact agriculture could be converted to intensive agricultural use (TEEB 2008).

HABITATS AND BIODIVERSITY

The use of pesticides particularly herbicides and synthetic fertilisers has increased dramatically over the past 60 years. In industrialised countries, farming practices have fundamentally changed. In the UK and many other places, mixed agriculture has been lost and farms have become increasingly specialised. Arable farming (field crops) and pastoral grassland are now largely separated as traditional crop rotation has been abandoned. In the British lowlands, field sizes have risen and field margins have shrunk. Harvesting has become more efficient and hedgerows have been lost. There has been a marked population decline of many species living on farmland (Boatmanetal. 2007).

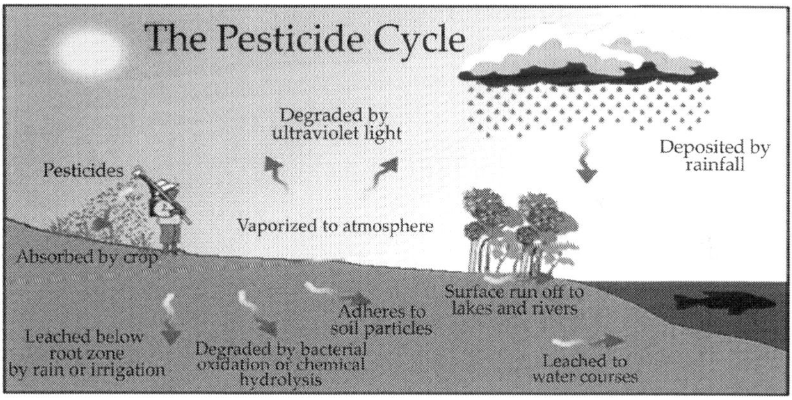

Figure 10.1: The Pesticide Cycle on how pesticide affects the environment.

Worldwide humans are estimated to use about 20% of the net primary production (plant organic matter produced in photosynthesis). In South America and Africa, humans use about 6% and 12%, respectively, of the regional net primary production, while the fraction consumed by humans is 72% in western Europe and 80% in south central Asia (Imhoff et al.,2004).The proportion of plant organic matter consumed by humans varies enormously between regions. For example, consumption of regionally-produced plant matter per inhabitant is nearly five times higher in North America than in south central Asia.Changes of habitat and biodiversity have been due both to changing climate and people's increasing use of plant and animal resources.

IMPACT OF PESTICIDES ON WILDLIFE POPULATIONS AND SPECIES DIVERSITY

Pesticides maybe advantageous to farmers but sometimes they can be are toxic to beneficial insects, birds, mammals, amphibians, or fish. Actually wildlife can be poisoned depending on the use of pesticide with high toxicity level because they tend to be persistent to the environment such as water-soluble pesticides may pollute surface waters), and they are also dose dependent such as the quantity applied, frequency, timing and method of application such spraying which can either be fine spray is prone to drift, weather, vegetation structure, and soil type. Insecticides, rodenticides, fungicides to control molds that may affect the seed and the more toxic herbicides threaten exposed wildlife.Over the past 40 years, the use of highly toxic carbamate and organophosphate has strongly increased. In the south, organochlorines such as endosulfan, highly persistent in the environment, are still used on a large scale. With habitat change, pesticide poisoning can cause major population decline which may threaten rare species.

Agricultural pesticides can reduce the abundance of weeds and insects which are important food sources for many species. Herbicides can change habitats by altering vegetation structure, ultimately leading to population decline. Fungicide use has also allowed farmers to stop growing 'break crops' like grass or roots. This has led to the decline of some arable weeds (Boatmanetal 2007).

In Canada, losses among 62 imperilled species were significantly more closely related to rates of pesticide use than to agricultural area in a region. Species loss was highest in areas with intensive agriculture (aerial spraying).The authors concluded that either pesticides, or other features of intensive agriculture linked to pesticide use in Canada, played a major part in the decline of imperilled species (Gibbsetal2009).

- Pesticides affect wildlife directly and indirectly via food sources and habitats such as:

- Wildlife poisoning by highly toxic insecticides, rodenticides, fungicides (on treated seed) and toxic herbicides can cause major population decline;
- Pesticides accumulating in the food chain, particularly those which cause endocrine disruption, pose a long-term risk to mammals, birds, amphibians, and fish;
- Broad-spectrum insecticides and herbicides reduce food sources for birds and mammals. This can produce a substantial decline in rare species populations and
- By changing vegetation structure, herbicides can render habitats unsuitable for certain species. This threatens insects, farmland birds, and mammals.

Bird species decline owing to pesticides

In western Europe the number of farmland birds is now just half that of 1980, even among formerly abundant species.While average populations of all common and forest birds declined by about 10% in Europe between 1980 and 2006, populations of farmland birds have fallen by 48%.This figure is based on surveys in 21 EU countries (EBCC 2008). Forest birds have declined less than have specialist birds living on farmland. A recent survey found that in the USA one in three bird species is endangered, threatened, or of conservation concern (NABCI etal 2009). Forty percent of grassland and arid-land birds are affected by population decline. Populations of raptors and other birds recovered after DDT and other toxic pesticides were banned in Europe. In North America between 1980 and 1999, populations of grassland species declined more than species living in shrubland. In 78% of species there was at least one association between population trend and change in agricultural land-use, and for most species this factor accounted for 25-30% of the variation in trend among states (Murphy 2003).

In Europe, the population decline among farmland birds was far greater in countries with more intensive agriculture, and in a statistical analysis 'cereal yield' explained over 30% of the trend in population change (Donaldetal2001). The authors of this study predicted that introducing EU agricultural policy into accession countries will result in a major decline in key bird populations.This occurred in the German state of Saxony-Anhalt. After 1990, farming in this region shifted from rotational cultivation (eg root-crops) to oilseed rape and winter cereals, which led to a reduction in grassland area and increased insecticide and herbicide use. In the same period, red kite (Milvus milvus) numbers fell by 50% from over 40 nesting pairs to about 20 pairs per 100 km2 (Nicolaietal 2009).

Important Bird Areas (IBAs) include agricultural areas with important bird populations. Although IBAs are appointed as priority conservation sites, they have no official protected status (Heath & Evans 2000). Agricultural expansion and intensification threaten half of IBAs in Africa and one-third in Europe. It is estimated that worldwide bird populations have declined by 20% to 25% since preagricultural times. Altogether, 1,211 bird species (12% of the total) are considered globally threatened, and 86% of these are threatened by habitat destruction or degradation. Birds species are globally threatened (as shown in figure11.1) wherein the primary pressure is chemical pollution, including fertilisers, pesticides and heavy metals entering surface water and the terrestrial environment (BLI 2004).

Figure 10.2: Migatory birds tend to look for another sanctuary due to chemical pollution.

BIRD POISONINGS CAUSED BY PESTICIDES

In the UK, the volume of seeds eaten by many bird species is large enough to pose a potential risk if the seeds are treated with one of the more toxic fungicides (Prosser & Hart 2005). Organophosphate insecticides, including disulfoton, fenthion, and parathion are highly toxic to birds. These have frequently poisonined raptors foraging in fields (Mineau 1999).Field studies have led to the conclusion that given the usual amounts of insecticide used, "direct mortality of exposed birds is both inevitable and relatively frequent with a large number of insecticides currently registered" (Mineau 2005). In the USA, some 50 pesticides have killed songbirds, gamebirds, raptors, seabirds and shorebirds (BLI 2004). In a small area of the Argentine pampas, monocrotophos, an organophosphate, has killed 6,000 Swainson's hawks.Worldwide, over 100,000 bird deaths caused by this chemical have been documented (Hooper 2002).

The number of birds found killed by pesticides, in the UK was at least 60 in 2006, and 55 in 2007. Pesticides were investigated as a possible cause of death in another 90 cases (80, in 2007). The affected species included buzzard, red kite, raven, crow, peregrine falcon, golden eagle, gull, barn owl, tawny owl, magpie, pheasant, rook, marsh harrier, dove, jackdaw, and chaffinch (PSD & Defra 2007; ACP 2008). The following pesticides were identified as a cause of fatal bird poisoning: carbamates (aldicarb, bendiocarb, carbofuran), organophosphates (chlorpyrifos, diazinon, isofenphos, malathion, mevinphos, phorate), anticoagulant rodenticides (bromadiolone, brodifacoum, difenacoum), and alphachloralose.

In 2005, from 20 dead barn owls and ten kestrels that contained one or more anticoagulant rodenticides, six barn owls and five kestrels had residues in the potentially lethal range. It was concluded that rodenticides may have contributed to the death of one barn owl and two kestrels, based on the circumstances of death and examination of carcasses (Walker et al.,2007). Residues in five of 23 red kites found dead would be potentially lethal to barn owls, while 17 of these had residues of at least one rodenticide, and ten had residues from two or three rodenticides (Walkeret al., 2008). From dead tawny owls collected under the Predatory Bird Monitoring Scheme, 20% (and 33% of owl livers) contained residues of one or more rodenticides (Walker et al., 2008).

NEGATIVE IMPACTS OF PESTICIDES ON FOOD SOURCES OF BIRDS

Herbicides and avermectin residues (used as worming livestock agents) affect birds indirectly by reducing food abundance (Vickery et al.,2001). Lower availability of key invertebrates and seed food for farmland birds in northern Europe was likely due to insecticides and herbicides, intensification and specialisation of farmland, loss of field margins, and ploughing (Wilson et al., 1999). Insecticides generally had a negative effect on yellowhammer when spraying occurred during the breeding season. Spraying at this time may cause more damage than repeated use throughout the year (Morris et al., 2005). Spraying insecticides within 20 days of hatching led to smaller brood size of yellowhammer, lower mean weight of skylark chicks, and lower survival of corn bunting chicks (Boatman et al., 2004).More frequent spraying of insecticides, herbicides, or fungicides was linked to a considerably smaller abundance of food invertebrates.This resulted in lower breeding success of corn buntings and may have contributed to their decline (Brickle, 2000). In Sussex, herbicides were a major cause of the decline of grey partridge populations by removing weeds which are important insect hosts (GCT 2004).

Pesticide use trends (measured by the percentage of treated area) was linked to periods of rapid bird decline (Campbell & Cooke, 1997).Bird species at risk from indirect effects caused by pesticides in the UK include grey partridge, corn bunting, yellowhammer, red-backed shrike, skylark, tree sparrow, and yellow wagtail (CSL et al.,2005).The main causes of farmland bird decline have been (1) pesticides and weed-control with herbicides, particularly, (2) change from spring-sown to autumn-sown cereals, (3) drainage and intensified management of grassland, and (4) increasedcattle or sheep density (Newton 2004).Sublethal effects on the nervous system can cause changes in behaviour. In an orchard, parent birds made fewer feeding trips after azinphos-methyl, an organophosphate, had been sprayed (Bishop et al., 2000).

- Bird populations are directly affected by poisoning from organophosphate or carbamate insecticides and anticoagulant rodenticides. Sublethal poisoning of birds by organophosphates can lead to detrimental changes in behaviour.

- Broad-spectrum herbicides threaten rare and endangered bird species by reducing the abundance of weeds (eaten by birds) and insects hosted by weeds. Insecticides reduce the number of insects which are important food sources for birds.

RISK TO MAMMALS OF HAZARDOUS PESTICIDES

Pesticides and other chemicals have caused population declines in Britain's wild mammals. Mostly bats and rodents (and 38% of species) were affected (Harris,1995). Certain pesticides can gradually accumulate in the food chain.This is of concern to vertebrates, particularly species in higher orders and top predators such as mammals or raptors. Anticoagulant rodenticides are highly toxic and some can bioaccumulate. Non-target predatory mammals (eg dogs and foxes) and raptors frequently suffer'secondary poisoning'by eating rats or mice which are poisoned by rodenticides.In France, foxes were poisoned by residues of bromadiolone in prey tissue (Berny et al., 1997). In the UK, following rat control with rodenticides, local wood mice, bank vole, and field vole populations of declined significantly (Brakes & Smith ,2005).

At least 25-35% of small mammal predators (polecats, stoats, and weasels) sampled had been exposed to rodenticides and this may be an underestimate (Shore et al.,1999).However, it is not wellknown how frequently rodenticides cause secondary mammal poisoning, and what the impact on their populations might be.

Herbicide use can affect mammals such as the common shrew, wood mouse and badger by removing plant food sources and changing the microclimate (Hole et al.,2005). Hares prefer a more diverse habitat. They are thus likely to benefit from increased fallow land (Smith et al.,2004). On organic farms, foraging activity by bats was significantly higher than on conventional farms, which may be due to a larger abundance of prey insects (Wickramasinghe et al.,2004). Less intensive farming systems may help to reverse bat decline.

- Anticoagulant rodenticides often indirectly poison predatory mammals and raptors.

- Herbicides can cause changes in vegetation and habitat which threaten mammals, while insecticides may reduce the availability of important food insects.

Figure 10.3: The bears are threatened to go hungry due to scarcity of nourishment especially food insects.

IMPACT OF PESTICIDES ON BUTTERFLIES, BEES, AND NATURAL ENEMIES

Broad-spectrum insecticides (eg carbamates, organophosphates and pyrethroids) can cause population declines of beneficial insects such as bees, spiders, or beetles. Many of these species play an important role in the food web or as natural enemies of pest insects. Since 1970, insect numbers in cereal fields in Sussex have dropped by half (GCT 2004). Numbers of bugs, spiders and beetles were considerably higher in untreated fields (Moreby & Southway, 1999). On British organic farms, numbers and species richness of butterflies was greater than on conventional farms (Feber et al., 2007). The number of carabid beetles and spiders was usually higher on organic farms. Conventional management practice appeared to affect natural enemies far more than other insects or target pests (Bengtsson e tal.,2005). Moths were considerably more abundant on organic farms and species richness was higher (Wickramasinghe et al.,2004). In arable fields, insecticide use was an important factor influencing communities of epigeic spiders (Drapela et al.,2008).On sites with an increased pesticide input, communities of bugs, wild bees, and spiders were more uniform, indicating less exchange between communities in areas with intensive agriculture (Dormann et al., 2008).

Bees perform essential pollination. Honey bees are under pressure from parasitic mites, viral diseases, habitat loss and pesticides. Intensified agricultural practices, habitat loss, and agrochemicals are considered to be among the chief environmental threats to Europe's honey and wild bees.Agricultural policy must reduce these pressures to ensure adequate pollinator populations (Kuldnaetal2009). On organic farms in the USA, near natural habitat, diverse native wild bee communities provided full pollination services, while diversity and numbers of native bees were greatly reduced on other farms (Kremenetal2002).In the UK, of the 95 incidents of bee poisoning (where the cause could be identified) between 1995 and 2001, organophosphates caused 42%, carbamates 29%, and pyrethroids 14% of cases (Fletcher & Barnett, 2003).In the last decade in the UK, insecticides which poisoned bee colonies included bendiocarb (a carbamate) and three pyrethroids: cypermethrin, deltamethrin and permethrin (PSD 2001-7). Synergistic effects between pyrethroids and EBI fungicides (imidazole or triazole fungicides) can increase the risk to honeybees (Pilling & Jepson, 2006).

Clothianidin, and to a lesser extent, imidacloprid are highly toxic to bumble bees and other wild bees (Scott-Dupree et al.,2009).These two neonicotinoid insecticides are used to treat corn and sunflower seeds. In 2008, clothianidin caused many bee poisonings and colony deaths in southern Germany (Spiegel 2008).The product has since been withdrawn. When imidacloprid-treated seed is grown, a large enough amount can enter the environment to poison bees (Greatti et al.,2003). Residues of imidacloprid in maize pollen grown from treated seed can be a high risk to bees owing to sublethal effects (Bonmatin et al.,2005). Even at low doses of imidacloprid, bees' foraging behavior was negatively affected (Yangetal2008).Exposure to low doses of imidacloprid over a longer period led to reduced learning capacity among bees (Decourtye et al.,2003).In alfalfa, imidacloprid affected the number and species diversity in communities of arthropods (natural enemies such as spiders) more strongly than among target pest insects (Liu et al.,2008).

Imidacloprid has been banned in France. Field margins without use of pesticides (herbicides in particular) had a positive effect on the number of Lepidoptera (such as moths or butterflies), bugs, and staphylinid beetles at the edges of arable fields (Frampton et al.,2007).In organic plots, average numbers of spiders and carabid or staphylinid beetles were almost twice as high as those in conventional plots (Mäder et al.,2002).

- Pesticides which are highly toxic to bees, bumblebees and other beneficial insects: carbamates (eg aldicarb, benomyl, carbofuran, methiocarb), organophosphates (eg chlorpyrifos, diazinon, dimethoate, fenitrothion), pyrethroids (eg cyfluthrin, cyhalothrin), and neonicotinoids (imidacloprid, thiamethoxam, clothianidin).
- Recently, clothianidin used in seed treatments have caused widespread bee poisoning. Imidacloprid residues in plants can negatively alter bee behaviour.

PESTICIDES AFFECTING AMPHIBIANS AND AQUATIC SPECIES

One-third of 6,000 amphibian species worldwide are threatened. Besides habitat loss, overexploitation or introduced species, amphibians are affected by the pollution of surface waters with fertilisers and pesticides from agriculture (IUCN 2009). In the USA, spray drift of hexazinone, a triazine herbicide, was considered "likely to adversely affect" the endangered red-legged California frog and its habitat (US EPA 2008). Atrazine is moderately toxic to some fish species.It can indirectly affect aquatic ecosystems by damaging aquatic plants. A review concluded that further study is needed on the potential hormonal effects of atrazine on frogs or fish (US EPA 2006). In Europe, the authorisation for atrazine has been withdrawn due to health and environmental risks (EC 2003).Urea herbicides such as isoproturon and diuron often contaminate rivers, lakes, and groundwater. Most breakdown products of diuron were more toxic to cellular microorganisms than the parent compound (Bonnet et al.,2007). Fungicides based on copper are highly toxic to aquatic organisms. In fish and some other aquatic organisms, the risk of copper accumulating may be high (EFSA 2008).The EU aims eventually to eliminate copper in organic vineyards and apple orchards (REPCO 2007).

A major study investigating amphibian communities in the USA found that, among other factors, agricultural fields near surface water and pesticides (at sufficiently high concentrations to affect insects or plants) will harm amphibian species richness.In particular, if water contains herbicides at concentrations which substantially reduce populations of aquatic plants this is likely to be associated with low numbers of amphibians relative to predator populations, and increased numbers of trematode parasites in amphibians (Beaslyetal2002).Parasitic nematodes were more abundant in agricultural wetlands during the growing season. Agricultural activity may intensify the infection of frogs by harmful nematodes (King et al.,2008). Atrazine suppressed the immune system of tiger salamanders by causing reduced numbers of white blood cells. The rate of infection by pathogenic viruses was higher in salamanders which were exposed to atrazine (Forson & Storfer 2006).In field studies, atrazine affected the immune system of tadpoles of northern leopard frogs, a declining species. Atrazine and phosphate fertiliser were the main factors linked to numbers of larval trematodes in frogs (Rohr et al.,2008). In California, tadpoles of the Pacific treefrog in areas with a poor population status had reduced levels of the enzyme cholinesterase, indicating exposure to organophosphates and/or carbamates (Sparling et al., 2001). Endosulfan was highly toxic to the declining yellow-legged frog. The insecticides chlorpyrifos and endosulfan have the potential to cause serious damage to amphibians at concentrations occurring in the environment under normal conditions of use (Sparling & Feller 2009). In laboratory tests, the survival of juvenile Great Plains toads and New Mexico spadefoot toads was reduced after exposure to certain formulations of the herbicides glufosinate and glyphosate (Dinehart et al.,2009).

Figure 10.4: Aquatic species of yellow frog has a higher risk of being eradicated in its existence.

Circumstances for pesticide use which present a high risk to communities of aquatic species include spray drift for insecticides and run-off from fields for herbicides (Verro et al.,2009). However, in a study of the risks of 261 pesticides to aquatic ecosystems in field ditches, about 95% of the predicted risk was caused by only seven pesticides (De Zwart 2005).More selective pesticides (with no or minimal impact on non-target organisms) should clearly be preferred. Surface water is frequently contaminated with insecticides through normal use at levels above those known to affect fish and aquatic invertebrates such as daphnia or shrimp.For example, this was observed for levels of azinphos-methyl, chlorpyrifos, and endosulfan in the aquatic environment (Schulz,2004). Similarly, chlorpyrifos and endosulfan were rated as 'chemicals of potential ecological concern'. An assessment concluded that adverse effects of endosulfan on fish and invertebrates are a concern when this insecticide is used near aquatic ecosystems (Carriger & Rand, 2008). In field tests, the insecticide carbaryl affected the composition of an aquatic community of amphibians and insects by changing colonisation of pools and numbers of eggs laid (Vonesch & Klaus ,2009).

- Insecticides and herbicides in surface waters (from spray-drift or run-off) can alter the species composition of aquatic communities and affect fish and invertebrates.
- Insecticides (organophosphates, carbamates) have toxic effects on the nervous systems of amphibians which may alter their

behaviour. Herbicides (eg atrazine) can impair the immune system of frog tadpoles, which can make amphibians more susceptible to harmful parasitic nematodes. Indirect effects can be fatal.

- Urea herbicides such as diuron frequently contaminate surface and ground water.Copper-based fungicides are highly toxic to fish and have a potential to accumulate.

EFFECT OF PESTICIDES ON PLANT COMMUNITIES

In recent decades, the use of herbicides has dramatically increased. Today, some non-crop plants (or 'weeds') are threatened with extinction in Britain (Preston,2002). Although the total volume of herbicides applied in the UK decreased slightly between 1990 and 2006, the herbicide-treated area increased by 38% (Fera,2009). Diversity of wild plants in agricultural fields and field margins is declining, especially in infertile grassland and hedge bottoms. A slight increase in plant diversity in arable fields in 1998 may have been due to the introduction of set-aside (Defra,2008). The number of plants providing food for butterfly caterpillars decreased in Britain between 1998 and 2007. Field margins created within agri-environment schemes supported a higher number of plant species than crop areas, but plant cover and species richness are still low (on average 11 species per plot and 21% cover) when compared to other habitats (eg horticultural land) and set-aside (CS 2007). By providing an unsprayed field margin at least three metres wide, the diversity and number of arable plants and insects hosted by them increased substantially (De Snoo,1999).Over five years, harmful weed cover did not increase in field margins (Mustersetal2009). On European farms using IPM methods (and an average of half the herbicides), the seed bank of weeds in soil doubled in autumn-sown crops.This was considered acceptable, while in spring-sown crops the seed bank increased more than threefold for certain weeds (EN 2005). In lowland areas in England, species diversity and abundance of plants, birds, bats, invertebrates, and plants were typically higher on organic farms than conventional ones. Positive effects were strongest for plants. It was estimated that organic fields held up to twice as many plant species and, on average, a weed cover twice as large (Fuller et al.,2005).

Flora in Britain is changing as arable plants such as the corn marigold introduced long ago are decreasing. This could be due to intensive agriculture and a decline in mixed farming (Preston, 2009).

Some herbicides are highly toxic to plants at very low doses, eg sulfonylureas, sulfonamides and imidazolinones. Tribenuron-methyl affected growth of algae and activity of microalgae at very low concentrations (Nystrom etal 1999). In a study on the effects of sulfonylurea herbicides on phytoplankton it was concluded that these herbicides present a potential hazard to aquatic systems even at low environmental concentrations (Sabater et al.,2002). Sulfonylureas have replaced other herbicides which are more toxic to animals. In potatoes, sulfometuron-methyl caused major yield losses even when used at rates below the recommended dose (Pfleegeretal2008). Experts have warned that the wide-spread use of sulfonylureas "could have a devastating impact on the productivity of nontarget crops and the makeup of natural plant communities and wildlife food chains", (Fletcher et al.,1993). Hexazinone is a persistent triazine with high leachability (Footprint 2009). In the USA, at all application rates the EPA's levels of concern for aquatic and terrestrial non-target plants were exceeded. Aquatic ecosystems within or next to hexazinone-treated areas could be altered by the effects on aquatic plants (US EPA 1994). Other triazines affect aquatic plants similarly, eg terbuthylazine and atrazine. In field tests, the herbicide glyphosate altered the composition of freshwater microbial communities by decreasing the abundance of microbial phytoplankton and increasing cyanobacteria (Pérez et al.,2007).

- Many plants which were previously common on British farmland are declining owing to the abandonment of mixed farming and increased herbicide use.
- Large-scale use of sulfonylurea herbicides, and presumably also sulfonamides and imidazolinones, poses a risk to non-target plants, algae, and ecosystems.
- Triazine herbicides may present a risk to non-target and aquatic plants.

Are pesticides diminishing soil fertility?

In South Africa, the feeding activity of soil organisms was higher in soil from organic vineyards than from conventionally treated sites (Reinecke et al.,2008). The number of earthworms was 1.3-3.2 times higher in organic compared to conventional plots, and the length of plant roots colonised by mycorrhizae was 40% higher in organic than in conventional systems (Mäder et al.,2002).Triclopyr, a herbicide, caused a major reduction in the growth of mycorrhizae at levated soil levels (Chakravarty, 1987).

The sulfonylurea herbicides metsulfuron and (to a lesser extent) chlorsulfuron caused a reduction in the growth of Pseudomonas soil bacteria (Boldt & Jacobsen 2006). In laboratory tests, a combination of two sulfonylurea herbicides, bensulfuron-methyl (B) and metsulfuron-methyl, caused a considerable reduction in soil microbial biomass over the first 15 days (El-Ghamry et al.,2001). In bacterial communities in soil, bromoxynile (a nitrile herbicide) caused major changes in species composition and diversity. Bromoxynile inhibited the growth of bacteria capable of degrading chemicals in soil (Baxteretal2008).Also captan (a fungicide) and the herbicide glyphosate caused a shift among species in bacterial communities in soil (Widenfalk et al.,2008). Certain organophosphate insecticides (eg dimethoate) can decrease the activity and biomass of soil microorganisms, while others (such as fosthiazate) may actually result in an increase in microbial biomass (Eisenhauer et al.,2009).How pesticides affect long-term soil fertility is not well understood as this depends on many factors.

- Pesticides affect earthworms, symbiotic mycorrhizae, and other organisms in soil.
- Composition and activity of bacterial communities can be changed by pesticides.

Figure 10.5: Spraying using backpack in a vineyard of grapes orchard.

POLICIES AND METHODS FOR BIODIVERSITY CONSERVATION

Actually in Europe, national policies set targets for biodiversity conservation (EC 2009).The Convention on Biological Diversity provides national strategies and action plans for conserving species at national level. These include the establishment of national targets. For example, the UK Biodiversity Action Plan (BAP) currently lists 1,150 species and 65 habitats with a priority for conservation. In 2002, of 78 farmland priority species, 39% were declining, 21% had unknown or unclear status, 18% were stable, 15% on the increase, and 7% had been lost. From the total area of one million hectares of nationally-important wildlife sites in the UK in 2003, about 380,000 hectares, or 38%, were in an unfavourable condition owing mainly to agriculture. Only 47% of important wildlife sites on farmland were in a favourable condition (Defra, 2003).One of the BAP's targets is to reverse the decline in farmland birds in Britain by 2020. In winter, farmland bird density is much higher on stubble (rotational set-aside) than on cereal fields. However, EU policy recently changed and set-aside is no longer compulsory. Ornithologists have warned that this could have serious negative impacts on farmland biodiversity across the EU (Gillings et al.,2009). Maintaining an appropriate population of weed species to support farmland wildlife is a challenge.It may be achieved by providing conservation headlands, by developing much more selective herbicides, and through their selective use (IACR 2001). In England between 1978 and 1990, plant diversity on arable land was declining. Between 1998 and 2007, plant diversity in main plots increased by 36%.This was due to increases in the area of set-aside or fallow land, driven by agrienvironment schemes (CS 2009). On plots with reduced herbicide input, farmland birds used winter cereal stubble more often than on conventional plots (Bradbury et al.,2008). To reverse the decline of birds, farming needs to change substantially and incorporate appropriate practices (Newton 2004). Species diversity is usually higher in non-crop areas than in grassland or fields. Unsprayed headlands of fields contain rare weed species and the highest diversity of invertebrates.As a feeding area for birds, the headland is most important. Weeds on the ground provide refuge and host many food insects. Using more selective herbicides in winter cereals could benefit farmland bird species which feed their chicks weed seeds eg linnets or finches (Moreby & Southway 1999). In the EU on arable or mixed farms which use integrated management practices, on average the use of herbicides was reduced by 43%, use of insecticides or molluscicides by 55%, and fungicide use was 50% lower when compared to conventional farms. On farms using IPM, the number of arthropods (such as beetles, spiders, springtails or sawflies), plants and earthworms increased significantly. Similar positive effects were observed for soil organisms, birds, and mammals such as wood mice (EN 2005).

National systems for pesticide authorisation aim to limit of the harm pesticides inflict on non-target species. But measures for reducing risks from pesticides are still being developed. Regulatory controls alone will not eliminate the impact on non-target species. Additional initiatives are needed which mitigate the effects of pesticides on biodiversity. The EU's Sixth Environment Action Programme identified biodiversity conservation as a high priority (EC 2002). Areas protected under the Birds and Habitats Directives are connected in the'Natura 2000'network. The proposed strategy on sustainable pesticide use in the EU aims to minimise risks to health and the environment from pesticides.Member states must eliminate or reduce the use of pesticides as far as possible in Natura 2000 sites, and promote farming with low pesticide input, particularly integrated pest management (IPM), and establish the necessary conditions for implementing IPM techniques (EC 2009).

POLICIES AND METHODS FOR BIODIVERSITY CONSERVATION

Agri-environment schemes in the EU provide payments to farmers taking measures to preserve the environment and countryside. But spending on these measures are to date marginal. Farmers who practise less intensive farming and who conserve nature need to be rewarded (Donald et al.,2002). The UN Convention on Biological Diversity requires that countries develop national strategies and action plans for conserving biodiversity, and define national or sub-national targets and indicators (CBD 2008). The number and quality of conservation targets vary strongly between countries (EPBRS 2009).Incentive schemes should be continuously evaluated and adapted (Berendse, 2004). To assess the effectiveness of measures for conserving threatened species such as farmland birds, quantitative and measurable goals are needed, plus monitoring of species (Donald et al.,2007). Extensive farming may need to stretch over larger connected areas to achieve substantial gains (Whittington 2007). Organic farms, together with agri-environment schemes to some degree, have positive effects on the diversity of plants and beetles in the EU, while bird species were not significantly more diverse. This may be due to widespread chemical pollution. So a shift towards farming with minimal pesticide use over large areas is urgently needed (Geiger et al.,2010). Corn, sugar cane and palm oil are increasingly being used to produce biofuels.These crop plants are linked to a high input of pesticide and fertiliser.Their use as biofuels is threatening biodiversity. Corn-based bioethanol is the worst among the alternatives currently available, so alternatives to corn as a biofuel source must be urgently pursued (Groom et al.,2008).In the EU, to produce energy from plant biomass in an environmentally-friendly way, the proportion of environmentally-orientated farming would need to increase to about 30 % of the Utilised Agricultural Area in most Member States by 2030, except for the most densely-populated countries (EEA 2007).

The need for a biodiversity rescue plan

The UN Convention on Biological Diversity requires the EU's 27 Member States to develop national policies to set biodiversity conservation targets. Not all Member States are equally ambitious, meaning that the 2010 objectives to halt further biodiversity loss need a new quantitative rescue plan for 2020, setting clear quantitative and qualitative targets, timetables and requiring ambitious monitoring. They also need to ensure coherence and better targeting, on these and a number of other EU policies (for sensitive'Natura 2000'areas and water), the establishment of new EU policies (on soil and bio-waste). However, the success of the biodiversity rescue plan will also to a large extent depend on the EU's implementation of the new'Regulation on the Placing of Plant Protection Products on the Market', as well as on how seriously member states implement the new framework directive on the sustainable use of pesticides. An important tool would be for member states to use this new opportunity to set dependency/use pesticide reduction targets and clear timetables.

A biodiversity rescue plan also needs to be accompanied by further reform of the EU's Common Agricultural Policy (CAP), departing from the current model where farmers receive income support for up-keep of their land into a model where farmers receive funding to provide public benefits, which includes paying farmers to use sustainable agricultural practices based on prevention first, also called integrated production, whereby the more farmers provide environmental and health services, the greater the public funding they receive.

In the International Year of Biodiversity 2010, we should fight together for reform of the CAP to encourage better agricultural practices. We should start by encouraging more mixed agriculture, crop rotation and pastoral grassland and lower field size. Even more so, we should encourage the development of practices such as bigger field margins and the re-establishment of hedgerows.We should put prevention first, in a dynamic system, encouraging front-runners who are willing to make environmental improvements, and incorporate a policy of making truly integrated agricultural production the basis of the post-2013 CAP .

Such an approach would be a step in the right direction in reversing the decline of birds, bees, bats, arthropods and earthworms, which thrive best in association with organic farming.It is also the best way to re-establish communities of different animal and plant species which perform vital functions within ecosystems, bringing higher diversity which tends to be more stable, and as a result will also help ensure greater long-term food security figure 11.5

Figure 10.6: A stable ecosystem which lead to a long term food security.

REFERENCES

1. Advisory Committee on Pesticides (ACP) – Environmental Panel, Pesticide poisoning of animals in 2007, York, UK 2008. http://www.pesticides.gov.uk/environment.asp?id=2600
2. Baxter J, and Cummings SP , The degradation of the herbicide bromoxynil and its impact on bacterial diversity in a top soil, Journal of Applied Microbiology 104(6): 1605-1616, 2008. http://www3.interscience.wiley.com/journal/120091862/
3. Beasly V,etal, Envrionmental factors that affect amphibian community structure and health as indicators of ecosystems, U.S. EPA, Washington D.C. 2002.
4. Berendse F ,etal, Declining biodiversity in agricultural Landscapes and the effectiveness of agri-environments chemes, Ambio 33(8): 499-502, 2004. http://www.bioone.org/doi/abs/10.1579/0044-7447-33.8.499 (abstract)
5. Berny PJ,etal, Field evidence of secondary poisoning of foxes (Vulpesvulpes) and buzzards (Buteo buteo) by bromadiolone, a 4-year survey, Chemosphere 35(8): 1817-1829, 1997. http://dx.doi.org/10.1016/S00456535(97)00242-7 (abstract)
6. BirdLife International (BLI), State of the world's birds, 2004. www.birdlife.org/sowb/ Bishop CA,etal, Effects of pesticide

spraying on chick growth, behavior, and parental care in tree swallows (Tachycinetabicolor) nesting in an apple orchard in Ontario, Canada, Environmental Toxicology and Chemistry 19(9): 2286-2297, 2000. http://dx.doi.org/10.1897/1551-5028(2000)019 <2286:EOPSOC>2.3.CO;2 (abstract)

7. Boatman ND,etal, Impacts of agricultural change on farmland biodiversity in the UK, In: Hester RE, and Harrison RM (eds), Biodiversity under threat, RSC Publishing, Cambridge, UK 2007, pp. 1-32.
8. Boatman ND,etal, Evidence for the indirect effects of pesticides on farmland birds, Ibis 146(2): 131-143, 2004. http://www3.interscience.wiley.com/journal/118753474/abstract (abstract)
9. Bonnet J.-L.,etal, Assessment of the potential toxicity of herbicides and their degradation products to nontarget cells using two microorganisms, the bacteria Vibrio fischeri and the ciliateTetrahymena pyriformis, Environmental Toxicology 22(1): 78-91, 2007
10. Bonmatin JM,etal, Quantification of imidacloprid uptake in maize crops. Journal of Agricultural and Food Chemistry 53: 5336-5341, 2005. In: UNAF: Pesticides
11. Brakes CR, and Smith RH, Exposure of non-target small mammals to rodenticides: short-term effects, recovery and implications for secondary poisoning, Journal of Applied Ecology 42(1): 118-128, 2005. http://www3.interscience.wiley.com/cgi-bin/fulltext/118735198/HTMLSTART
12. Brickle NW,etal, Effects of agricultural intensification on the breeding success of corn buntingsMiliaria calandra, Journal of Applied Ecology 37(5): 742-755, 2000.
13. Carriger JF , and Rand GM, Aquatic risk assessment of pesticides in surface waters in and adjacent to the Everglades and Biscayne National Parks: I. Hazard assessment and problem formulation, Ecotoxicology 17(7): 660-679, 2008. http://www.springerlink.com/content/w48p7732t04833w3/
14. CBD, Ninth meeting of the Conference of the Parties to the Convention on Biological Diversity, Bonn, Germany 19-30 May 2008. http://www.cbd.int/decisions/cop/?m=cop-09

15. (CBS 1963): Columbia Broadcasting System, C.B.S. Reports:The Silent Spring of Rachel Carson, April 3, 1963, In: Lerner KL, and Lerner BW (eds), Environmental issues: essential primary sources, Detroit: Thomson Gale 2006.
16. Central Science Laboratory (CSL), Game Conservancy Trust, RSPB, and Department of Zoology of Oxford University, Assessing the indirect effects of pesticides on birds, Final report, 2005. http://randd.defra.gov.uk/Document.aspx?Document=PN0925_2486_FRP .pdf
17. Chakravarty P , and Sidhu S, Effect of glyphosate, hexazinone and triclopyr on in vitro growth of five species of ectomycorrhizal fungi, Forest Pathology 17(4-5): 204-210, 1987
18. Countryside Survey (CS). England results from 2007. Chapter 3: Enclosed farmland, 2009. http://www.countrysidesurvey.org.uk/eng_reports2007.html
19. Darwin C, and Wallace A, On the tendency of species to form varieties; and on the perpetuation of varieties and species by natural means of selection, Journal of the Proceedings of the Linnean Society of London, Zoology 3: 45-62, 1858.
20. Department of Environment, Food and Rural Affairs (Defra), Measuring the progress of the biodiversity strategy for England: baseline assessment, London 2003, amended in 2008. http://www.defra.gov.uk/environment/quality/biodiversity/documents/indicator/indicators031201.pdf
21. De Snoo GR, Unsprayed field margins: effects on environment, biodiversity and agricultural practice, Landscape and Urban Planning 46(1-3): 151-160, 1999. http://dx.doi.org/10.1016/S0169-2046(99)00039-0
22. De Zwart, Ecological effects of pesticide use in the Netherlands: Modeled and observed effects in the field ditch, Integrated Environmental Assessment and Management 1(2): 123-134, 2005. http://www.bioone.org/doi/abs/10.1897/IEAM_2004-015.1 (abstract)
23. Dinehart SK,etal, Toxicity of a glufosinate- and several glyphosate-based herbicides to juvenile amphibians from the Southern High Plains, USA, Science of the Total Environment 407(3): 1065-1071,

2009. http://dx.doi.org/10.1016/j.scitotenv.2008.10.010 (abstract)

24. Donald PF ,etal, International conservation policy delivers benefits for birds in Europe, Science 317(5839):8 10-813, 2007. http://dx.doi.org/10.1126/science.1146002 (abstract)
25. Donald PF ,etal, The Common Agricultural Policy, EU enlargement and the conservation of Europe'sf armland birds, Agriculture, Ecosystems & Environment 89(3): 167-182, 2002. http://dx.doi.org/10.1016/S0167-8809(01)00244-4 (abstract)
26. Dormann CF ,etal, Effects of landscape structure and land-use intensity on similarity of plant and animalc ommunities, Global Ecology and Biogeography 16(6): 774-787, 2007.
27. (EPBRS 2009): Mindicate, and U&W, Targets for biodiversity beyond 2010 – Review and cases in a European context, Stockholm 2009. http://www.epbrs.org/PDF/Background%20report%20 EPBRS.pdf
28. English Nature (EN), Integrated farming and biodiversity, Peterborough, UK 2005. http://naturalengland.etraderstores.com/NaturalEnglandShop/R634
29. European Bird Census Council (EBCC), European wild bird indicators, 2008 update. www.ebcc.info/index.php?ID=368
30. European Commission (EC), Development of guidance for establishing Integrated Pest Management (IPM) principles, 2009. http://ec.europa.eu/environment/ppps/pdf/final_report_ipm.pdf
31. European Commission (EC), Draft Guidance Document for establishing IPM principles (supplement), 2009a. http://ec.europa.eu/environment/ppps/pdf/draft_guidance_doc.pdf
32. European Commission (EC), A mid-term assessment of implementing the EC Biodiversity Action Plan, COM(2008) 864 final, Brussels 2008.
33. European Commission (EC), Proposal for a Directive of the European Parliament and of the Council establishing a framework for Community action to achieve a sustainable use of pesticides, Brussels 2006. http://ec.europa.eu/environment/ppps/pdf/com_2006_0373.pdf

34. European Environment Agency (EEA), Estimating the environmentally compatible bioenergy potential from agriculture, 2007. http://www.eea.europa.eu/publications/technical_report_2007_12
35. European Commission (EC), Review report for the active substance atrazine, Brussels 2003, http://ec.europa.eu/food/plant/protection/evaluation/existactive/list_atrazine.pdf
36. European Commission (EC), Sixth Environment Action Programme of the European Community 2002-2012. Brussels 2002. http://ec.europa.eu/environment/newprg/index.htm 1920
37. European Food Safety Authority (EFSA), Conclusion regarding the peer review of the pesticide risk assessment of the active substance Copper (I), copper (II) variants namely copper hydroxide, copper oxychloride, tribasic copper sulfate, copper (I) oxide, Bordeaux mixture, In: Summary of the EFSA Scientific Report 187, Parma, Italy 2008.
38. Fletcher M, and Barnett L, Bee pesticide poisoning incidents in the United Kingdom, Bulletin of Insectology 56: 141-145, 2003.
39. Food and Environment Research Agency UK (Fera), Pesticide Usage Statistics:Tables (select year and chemical group), 2009. http://pusstats.csl.gov.uk/index.cfm Forson DD, and Storfer A, Atrazine increases Ranavirus susceptibility in the tiger salamander (Ambystoma tigrinum), Ecological Applications 16(6): 2325-2332, 2006.
40. Game and Wildlife Conservation Trust (GCT), Sussex study: 34 years of change in farmland wildlife, 2004. http://www.gct.org.uk/text03.asp?PageId=182
41. Geiger F .,etal, Persistent negative effects of pesticides on biodiversity and biological control potential onE uropean farmland, Basic and Applied Ecology 11(2): 97-105, 2010.
42. Gibbs KE,etal, Human land use, agriculture, pesticides and losses of imperiled species, Diversity and Distributions 15(2): 242-253, 2009.
43. Groom MJ, Gray EM, and Townsend PA, Biofuels and biodiversity: Principles for creating better policies for biofuel production, Conservation Biology 22(3): 602-609, 2008.

44. Heath MF , and Evans MI (eds.), Important Bird Areas in Europe: Priority sites for conservation, Volume 2, BirdLife International, Cambridge, UK, 2000,
45. Hooper MJ, Swainson's hawks and monocrotophos, Texas 2002.
46. IOBC, Toolbox for organizations seeking IOBC endorsement. http://www.iobc.ch/toolbox.html
47. IOBC, Integrated Production: Principles and Technical Guidelines, 3rd edition 2004. http://www.iobc.ch/iobc_bas.pdf
48. IUCN, Getting the biodiversity targets right, Trondheim Conference on Biodiversity 2010. http://www.countdown2010.net/article/getting-the-biodiversity-targets-right-in-trondheim
49. IUCN, The Asian amphibian crisis, 2009. http://www.iucn.org/about/union /secretariat/offices/asia/regional_activities/asian_amphibian_crisis/
50. International Organisation for Biological and Integrated Control of Noxious Animals and Plants (IOBC), IOBC Tool Box, 2005.
51. Kuldna P ,etal, An application of DPSIR framework to identify issues of pollinator loss, EcologicalE conomics, 69(1): 32-42, 2009. http://dx.doi.org/10.1016/j.ecolecon.2009.01.005
52. Liu CZ, Wang G, andYan L, [Effects of imidacloprid on arthropod community structure and its dynamics in alfalfa field], [Chinese Journal of Applied Ecology], 18(10): 2379-2383, 2008. http://www.ncbi.nlm.nih.gov/pubmed/18163327
53. Mäder P ,etal, Soil fertility and biodiversity in organic farming, Science 296(5573): 1694-1697, 2002,h ttp://www.sciencemag.org/cgi/content/abstract/296/5573/1694
54. Mineau P ,etal, Patterns of bird species abundance in relation to granular insecticide use in the Canadianp rairies, Ecoscience 12(2): 267-278, 2005. http://www.bioone.org/doi/abs/10.2980/i1195-6860-12-2-267.1 (abstract)

11
CHAPTER

GENERAL PROPERTIES OF HERBICIDES

INTRODUCTION

Herbicides belong to a group of chemicals known as pesticides, which prevent, destroy, repel, or mitigate any pest. Herbicides are any chemical substance that is used to specifically kill plants. Other familiar pesticides are insecticides, rodenticides, and fungicides.

MODE OF ACTION

The application of herbicides will result in different effects or mode of action biochemical or physical mechanism by which it kills plants. Most herbicides disrupt the growth of plants by controlling the growth process or altering one or more of their metabolic processes. This function of herbicides in the cellular membranes of plants, allowing cellular contents to leak out, but do not directly disrupt other metabolic processes. In some instances there is type of plants or whole groups of plants are not susceptible to certain herbicides because they use different biochemical pathways or have slightly different enzymes. In return animals is more tolerant in the pesticide effect and suffer little or no effect from most herbicides sold today because these compounds can help in suppressing or affect processes exclusive to plants, like photosynthesis or production of aliphatic amino acids. An herbicide is often chosen for use based on its effectivity in inhibiting the growth of weeds. If one herbicide is ineffective, another herbicide with a a better performance or a different mode of action may provide better results. When and how an herbicide is applied may be determined by its mode of action.

"Pre-emergent" herbicides are those applied to the soil before the weed germinates, and either disrupt germination or kill the germinating seedling. "Post-emergent" herbicides are those that are applied directly to already established plants and/or soil. Some herbicides are effective both before ("pre-emergent") and after ("post-emergent") germination.

Some of the most common modes of action include:

- Auxin mimics (2,4-D, clopyralid, picloram, and triclopyr), which mimic the plant growth hormone auxin causing uncontrolled and disorganized growth in susceptible plant species; *f*
- Mitosis inhibitors (fosamine), which prevent re-budding in spring and new growth in summer (also known as dormancy enforcers); *f*
- Photosynthesis inhibitors (hexazinone), which block specific reactions in photosynthesis leading to cell breakdown; *f*
- Amino acid synthesis inhibitors (glyphosate, imazapyr and imazapic), which prevent the synthesis of amino acids required for construction of proteins; *f*
- Lipid biosynthesis inhibitors (fluazifop-p-butyl and sethoxydim), that prevent the synthesis of lipids required for growth and maintenance of cell membranes.

AUXIN MIMICS

Picloram, clopyralid, triclopyr, and 2,4-D are referred to as synthetic auxins. Auxin is a plant hormone that regulates growth in many plant tissues. Chemically, 2,4-D is classified as a phenoxy acetic acid, while picloram, clopyralid, and triclopyr are pyridine derivatives. When susceptible plants are treated with these herbicides, they exhibit symptoms that could be called 'auxin overdose', and eventually die as a result of increased rates of disorganized and uncontrolled growth.

In use since 1945, 2,4-D is one of the most studied herbicides in the world. It is known to affect many biochemical processes in plants, but it is still not clear which of the biochemical alterations 2,4-D and other auxin-mimic herbicides cause that is ultimately responsible for killing plants. It is possible that plants are weakened more or less equally by several of these disruptions with no one process being the most important.

The sequence of events following treatment with an auxin mimic herbicide differs from one species to another and depends on the age and physiological state of the individual plant. Marked changes in the permeability of the plant's cell wall or membrane can generally be detected within minutes of application. This change may lead to a rapid and sustained loss of H+ ions (protons) from the cell wall, which makes the wall more elastic, and often results in measurable cell growth within an hour. The loss of H+ ions may also lead to an accumulation of K+ ions in the stomatal guard cells, causing those cells to swell, increasing the size of the stomatal opening. The increased stomatal opening helps bring about a short-lived increase in photosynthesis, presumably because it allows higher concentrations of CO2 to reach the photosynthesizing cells inside the leaf. Other biochemical changes that occur within a day of treatment include a large increase in the concentration of soluble sugars and amino acids inside cells. This increase coincides with an increase in messenger RNA synthesis and a large increase in rates of protein synthesis. Treated plants also frequently produce ethylene, a gaseous plant hormone.

Grasses and other monocots are generally not susceptible to auxin-mimic herbicides. The reason for this selectivity is unclear because there are no apparent differences between the binding sites targeted by auxins in monocots and dicots. It may, however, be due to differences in vascular tissue structure or differences in ability to translocate or metabolize the herbicide (DiTomaso 1999).

MITOSIS INHIBITORS

Fosamine ammonium is another herbicide that acts as a plant growth regulator. It is sometimes referred to as a "dormancy enforcer," but the specific mechanism of action has not been identified, even though there is evidence that fosamine ammonium inhibits mitosis in susceptible plants. When applied to deciduous plants up to two months before leaf drop, the compound is absorbed with little or no apparent effect. The following spring however, the plants fail to leaf-out because bud development is either prevented or limited to spindly, miniature leaves. Plants often die as the season progresses because they cannot produce enough photosynthate to sustain themselves. A distinctive feature of this mode of action is that treated plants do not go through a "brown-out" phase, as is often seen after the application of other herbicides. Susceptible non-deciduous plants such as pines, die soon after application because they simply do not produce enough photosynthate.

PHOTOSYNTHESIS INHIBITORS

There are two types of photosynthesis inhibitors. Hexazinone is an example of the type that inhibits the transfer of electrons in photosystem II. It blocks electron transport from QA to QB in the chloroplast thylakoid membranes by binding to the D-1 protein at the QBbinding niche. The electrons blocked from passing through photosystem II are transferred through a series of reactions to other reactive toxic compounds. These compounds disrupt cell membranes and cause chloroplast swelling, membrane leakage, and ultimately cellular destruction.

Paraquat and diquat are examples of the second type of photosynthesis inhibitor. They accept electrons from Photosystem I, and after several cycles, generate hydroxyl radicals. These radicals are extremely reactive and readily destroy unsaturated lipids, including membrane fatty acids and chlorophyll. This destroys cell membrane integrity, so that cells and organelles "leak", leading to rapid leaf wilting and desiccation, and eventually to plant death (WSSA 1994).

Amino Acid Synthesis Inhibitors

Glyphosate and imazapyr kill plants by preventing the synthesis of certain amino acids produced by plants but not animals. Glyphosate blocks the action of the enzyme 5enolpyruvylshikimate-3-phosphate (EPSP) synthase, which inhibits the biosynthesis of certain aromatic amino acids such as phenylalanine, tyrosine, and tryptophan. These amino acids are required for protein synthesis, which, in turn, is necessary for plant growth and maintenance. Other biochemical processes such as carbohydrate translocation, can also be affected by these herbicides. Although these effects are considered secondary, they may be important in the total lethal action of glyphosate.

Imazapyr, another amino acid synthesis inhibitor, kills plants by inhibiting the production of the branched-chain aliphatic amino acids (valine, leucine, and isoleucine), which are required for DNA synthesis and cell growth. It does this by blocking acetohydroxy acid synthase (AHAS), also known as acetolactate synthase (ALS). Plants treated with imazapyr usually die slowly. The time it takes for a treated plant to die is most likely related to the amount of stored aliphatic amino acids available to the plant. AHAS (ALS) is widespread in plants but the biochemical pathway it catalyzes is not found in animals.

LIPID BIOSYNTHESIS INHIBITORS

Fluazifop-p-butyl and sethoxydim are both grass specific herbicides that inhibit the synthesis of enzymes required for lipid synthesis. Both inhibit acetyl CoA carboxylase, the enzyme responsible for catalyzing an early step in fatty acid synthesis. Nonsusceptible broadleaf species have a different binding site, rendering them immune. The inhibition of acetyl CoA carboxylase and the subsequent lack of lipid production leads to losses in cell membrane integrity, especially in regions of active growth such as meristems. Eventually shoot and rhizome growth ceases, and shoot meristems and rhizome buds begin to die back.

FORMULATIONS

A herbicide formulation is the total marketed product, and is typically available in forms that can be sprayed on as liquids or applied as dry solids. It includes the active ingredient(s), any additives that enhance herbicide effectiveness, stability, or ease of application such as surfactants and other adjuvants, and any other ingredients including solvents, carriers, or dyes. The application method and species to be treated will determine which formulation is best to use. In most cases, manufacturers produce formulations that make applications and handling simpler and safer. Some herbicides are available in forms that can reduce risk of exposure during mixing, such as pre-measured packets that dissolve in water, or as a liquid form already mixed with surfactant and dye (e.g., Pathfinder II®).

Sprayable/liquid formulations include:

1. *Water-soluble formulations:* soluble liquids (SL), soluble powders or packets (SP), and soluble granules (SG). Only a few herbicidal active ingredients readily dissolve in water. These products will not settle out or separate when mixed with water.
2. *Emulsifiable formulations* (oily liquids): emulsifiable concentrates (E or EC) and gels (GL). These products tend to be easy to handle and store, require little agitation, and will not settle out of solution. Disadvantages of these products are that most can be easily absorbed through the skin and the solvents they contain can cause the rubber and plastic parts of application equipment to deteriorate.
3. *Liquid suspensions* (L for liquid or F for flowable) that are dispersed in water include: suspension concentrates (SC), aqueous suspensions (AS), emulsions of waterdissolved herbicide in oil (EO), emulsions of an oil-dissolved herbicide in water (EW), micro-encapsulated formulations (ME), and capsule suspensions (CS). All these products consist of a particulate or liquid droplet active ingredient suspended ina liquid. They are easy to handle and apply, and rarely clog

nozzles. However, they can require agitation to keep the active ingredients from separating out.

4. *Dry solids that are suspended in water:* wettable powders (W or WP), waterdispersible granules (WDG, WG, DG), or dry flowables (DF). These formulations are some of the most widely used. The active ingredient is mixed with a fine particulate carrier, such as clay, to maintain suspension in water. These products tend to be inexpensive, easy to store, and are not as readily absorbed through the skin and eyes as ECs or other liquid formulations. These products, however, can be inhalation hazards during pouring and mixing. In addition, they require constant agitation to maintain suspension and they may be abrasive to application pumps and nozzles.

Dry formulations include:

1. Granules (G) – Granules consist of the active ingredient absorbed onto coarse particles of clay or other substance, and are most often used in soil applications. These formulations can persist for some time and may need to be incorporated into the soil.
2. Pellets (P) or tablets (TB) – Pellets are similar to granules but tend to be more uniform in size and shape.
3. Dusts (D) – A dust is a finely ground pesticide combined with an inert or inactive dry carrier. They can pose a drift or inhalation hazard.

Salts vs. Esters

Many herbicidally active compounds are acids that can be formulated as a salt or an ester for application. Once the compound enters the plant, the salt or ester cation is cleaved off allowing the parent acid (active ingredient) to be transported throughout the plant. When choosing between the salt or ester formulation, consider the following characteristics:

Salts

Most salts are highly water soluble, which reduces the need for emulsifiers or agitation to keep the compound suspended.

- Salts are not soluble in oil;
- Salts generally require a surfactant to facilitate penetration through the plant cuticle (waxy covering of leaves and stems);
- Salts are less volatile than esters and
- Salts can dissociate in water. In hard water the parent acid (i.e. the active ingredient) may bind with calcium and magnesium in the water, precipitate out, and be inactivated.

Esters

Esters can penetrate plant tissues more readily than salts, especially woody tissue *f* Esters generally are more toxic to plants than salts *f* Esters are not water soluble and require an emulsifying agent to remain suspended in water-based solvents *f* Esters have varying degrees of volatility

Adjuvants

An adjuvant is any material added to a pesticide mixture that facilitates mixing, application, or pesticide efficacy. An adjuvant enables an applicator to customize a formulation to be most effective in a particular situation. Adjuvants include surfactants, stickers, extenders, activators, compatibility agents, buffers and acidifiers, deposition aids, de-foaming agents, thickeners, and dyes.

Surfactants

Surfactants are the most important adjuvants. They are chemical compounds that facilitate the movement of the active herbicide ingredient into the plant. They may contain varying amounts of fatty acids that are capable of binding to two types of surfaces, such as oil and water. Some herbicide formulations come with a surfactant already added, in others, surfactants can be added prior to application. Whether a surfactant should be added will be determined by the type of herbicide being applied and the target plant. Read the label for recommendations of appropriate surfactants.

Mechanisms of Dissipation

Dissipation refers to the movement, degradation, or immobilization of an herbicide in the environment.

Degradation

Degradation occurs when an herbicide is decomposed to smaller component compounds, and eventually to CO_2, water, and salts through photochemical, chemical, or biological (microbial metabolism) reactions (Freed and Chiou, 1981). Biodegradation accounts for the greatest percentage of degradation for most herbicides (Freed and Chiou, 1981). When a single herbicide degrades, it usually yields several compounds ("metabolites"), each of which has its own chemical properties including toxicity, adsorption capacity, and resistance to degradation. Some metabolites are more toxic and/or persistent than the parent compound. In most cases, the nature of the metabolites are largely unknown.

Photodegradation

Photodegradation refers to decomposition by sunlight. Sunlight intensity varies with numerous factors including latitude, season, time of day, weather, pollution, and shading by soil, plants, litter, etc. Studies of the photodegradation of herbicides are often conducted using UV light exclusively, but there is some debate as to whether most UV light actually reaches the surface of the earth. Therefore, photodegradation rates determined in the laboratory may over-estimate the importance of this process in the field (Helling et al., 1971).

MICROBIAL DEGRADATION

Microbial degradation is decomposition through microbial metabolism. Different microbes can degrade different herbicides, and consequently, the rate of microbial degradation depends on the microbial community present in a given situation (Voos andGroffman 1997, McCall et al. ,1981). Soil conditions that maximize microbial degradation include warmth, moisture, and high organic content.

Herbicides may be microbially degraded via one of two routes. They may be metabolized directly when they serve as a source of carbon and energy (i.e. food) for microorganisms (Hutzinger, 1981), or they may be co-metabolized in conjunction with a naturally occurring food source that supports the microbes (Hutzinger, 1981). Herbicides that are co-metabolized do not provide enough energy and/or carbon to support the full rate of microbial metabolism on their own.

There is sometimes a lag time before microbial degradation proceeds. This may be because the populations of appropriate microbes or their supplies of necessary enzymes start small, and take time to build-up (Farmer and Aochi 1987, Kearney and Karns, 1960). If this lag time is long, other degradation processes may play more important roles in dissipation of the herbicide (Farmer and Aochi ,1987). Degradation rates of cometabolized herbicides tend to remain constant over time.

Chemical Decomposition

Chemical decomposition is degradation driven by chemical reactions, including hydrolyzation (reaction with hydrogen, usually in the form of water), oxidation (reaction with oxygen), and disassociation (loss of an ammonium or other chemical group from the parent molecule). The importance of these chemical reactions for herbicide degradation in the field is not clear (Helling et al. ,1971).

Immobilization/Adsorption

Herbicides may be immobilized by adsorption to soil particles or uptake by nonsusceptible plants. These processes isolate the herbicide and prevent it from moving in the environment, but both adsorption and uptake are reversible. In addition, adsorption can slow or prevent degradation mechanisms that permanently degrade the herbicide.

Adsorption refers to the binding of herbicide by soil particles, and rates are influenced by characteristics of the soil and of the herbicide. Adsorption is often dependent on the soil or water pH, which then determines the chemical structure of the herbicide in the environment. Adsorption generally increases with increasing soil organic content, clay content, and cation exchange capacity, and it decreases with increasing pH and temperature. Soil organic content is thought to be the best determinant of herbicide adsorption rates (Farmer & Aochi ,1987, Que Hee and Sutherland 1981, Helling et al., 1971). Adsorption is also related to the water solubility of an herbicide, with less soluble herbicides being more strongly adsorbed to soil particles (Helling et al. 1971). Solubility of herbicides in water generally decreases from salt to acid to ester formulations, but there are some exceptions. For example, glyphosate is highly water-soluble and has a strong adsorption capacity.

The availability of an herbicide for transport through the environment or for degradation is determined primarily by the adsorption/desorption process (WHO ,1984). Adsorptionto soil particles can stop or slow the rate of microbial metabolism significantly. In other cases, adsorption can facilitate chemical or biological degradation (Farmer & Aochi ,1987). Adsorption can change with time and, in most cases, is reversible (i.e. the herbicide can desorb from the soil or sediments and return to the soil solution or water column).

Movement and Volatilization through the environment occurs when herbicides are suspended in surface or subsurface runoff, volatilized during or after application, evaporated from soil and plant surfaces, or leached down into the soil. Although generally studied and discussed separately, these processes actually occur simultaneously and continuously in the environment (Hutzinger, 1981).

Volatilization occurs as the herbicide passes into the gaseous phase and moves about on the breeze. Volatilization most often occurs during application, but also can occur after the herbicide has been deposited on plants or the soil surface. The volatility of an herbicide is determined primarily by its molecular weight. Most highly volatile herbicides are no longer used.

Volatility generally increases with increasing temperature and soil moisture, and with decreasing clay and organic matter content (Helling et al., 1971). The use of a surfactant can change the volatility of a herbicide (Que Hee and Sutherland 1981). In extreme cases, losses due to volatilization can be up to 80 or 90% of the total herbicide applied (Taylor and Glotfelty, 1988). Of the herbicides described in detail in this handbook, only 2,4-D and triclopyr can present significant volatilization problems in the field (T. Lanini, pers. comm.).

BEHAVIOR IN THE ENVIRONMENT

The most important factor determining the fate of herbicide in the environment is its solubility in water (Hutzinger, 1981). Water-soluble herbicides generally have low adsorption capacities, and are consequently more mobile in the environment and more available for microbial metabolism and other degradation processes. Esters, in general, are relatively insoluble in water, adsorb quickly to soils, penetrate plant tissues readily, and are more volatile than salt and acid formulations (Que Hee & Sutherland, 1981).

Soils

An herbicide's persistence in soils is often described by its half-life (also known as the DT_{50}). The half-life is the time it takes for half of the herbicide applied to the soil to dissipate. The half-life gives only a rough estimate of the persistence of an herbicide since the half-life of a particular herbicide can vary significantly depending on soil characteristics, weather (especially temperature and soil moisture), and the vegetation at the site. Dissipation rates often change with time (Parker and Doxtader, 1983). For example, McCall et al. (1981) found that the rate of dissipation increased untilapproximately 20% of the applied herbicide remained, and then declines. Nonetheless, half-life values do provide a means of comparing the relative persistence of herbicides.

The distribution of an herbicide in the soil is determined primarily by the amount, type, and surface area of clays and organic matter in the soil, the amount and quality of soil moisture, and soil temperature and soil pH (Helling et al. 1971). Most natural soils have pH values between 5 and 8 wherein the rainfall and the amount of leaching that has occurred strongly influences these values. In wet areas and/or coarse soils, cations can be leached out, leaving the soil acidic. In arid and semi-arid regions, soils retain cations and are more alkaline. Acidic soils can also be found in bogs where organic acids lower the soil's pH.

Water

Water bodies can be contaminated by direct overspray, or when herbicides drift, volatilize, leach through soils to groundwater, or are carried in surface or subsurface runoff. Amounts of leaching and runoff are largely dependent on total rainfall the first few days after an application. Total losses to runoff generally do not exceed five to ten percent of the total applied, even following heavy rains (Taylor and Glotfelty, 1988). High soil adsorption capacity, low rates of application, and low rainfall reduce total runoff and contamination of local waterways (Bovey et al., 1978).

The behavior of an herbicide in water is dictated by its solubility in water. Salts and acids tend to remain dissolved in water until degraded through photolysis or hydrolysis. Esters will often adsorb to the suspended matter in water, and precipitate to the sediments. Once in the sediments, esters can remain adsorbed to soil particles or be degraded through microbial metabolism. Highly acidic or alkaline waters can chemically alter an herbicide and change its behavior in water. The average pH of surface waters is between five and nine (Hutzinger, 1981).

Birds and Mammals

A herbicide's toxicity is described by its LD_{50}, which is the dose received either orally (taken through the mouth) or dermally (absorbed through the skin) that kills half the population of study animals. The oral LD50s reported here were determined for adult male rats. The dermal LD_{50} is were determined for rabbits. The LD_{50} is typically reported in grams of herbicide per kilogram of animal body weight. LD_{50} is are determined under varying circumstances so comparisons between different herbicidesmay provide only a rough sense of their relative toxicities. Dermal LD_{50} values may be more meaningful to herbicide applicators because they are more likely to be exposed to herbicide through their skin rather than by oral ingestion. In any event, very few people, even among applicators, are exposed to herbicide doses as high as the LD_{50}.

The LD_{50} does not provide any information about chronic, long-term toxic effects that may result from exposure to lesser doses. Sublethal doses can lead to skin or eye irritation, headache, nausea, and, in more extreme cases, birth defects, genetic disorders, paralysis, cancer, and even death. Impurities derived from the formulation of the herbicide and the adjuvants added to the formulation may be more toxic than the herbicide compound itself, making it difficult to attribute increased risks of cancer or other effects directly to a herbicide (Ibrahim et al., 1991).

The most dramatic effects of herbicides on non-target plants and animals often result from the habitat alterations they cause by killing the targeted weeds. For example, loss of invasive riparian plants can cause changes in water temperature and clarity that can potentially impact the entire aquatic community, and the physical structure of the system through bank erosion. Removing a shrubby understory can make a habitat unsuitable for certain bird species and expose small mammals to predation.

Aquatic Species

A herbicide's toxicity to aquatic organisms is quantified with the LC_{50}, which is the concentration of herbicide in water required to kill half of the study animals. The LC_{50} is typically measured in micrograms of pesticide per liter of water.

In general, ester formulations are more dangerous for aquatic species than salt and acid formulations because ester formulations are lipophilic or known as fat-loving, and consequently, can pass through the skin and gills of aquatic species relatively easily. Ester formulations, additionally, are not water soluble, and are less likely to be diluted in aquatic systems.

Soil Microbes

Herbicides have varying effects on soil microbial populations depending on herbicide concentrations and the microbial species present. Low residue levels can enhance populations while higher levels can cause population declines. In many cases, studies are too short in duration to determine the true long-term impacts of herbicide use on soil microbes.

HUMAN TOXICOLOGY

When proper safety precautions are taken, human exposure to herbicides used in natural areas should be minimal. Properly fitted personal protection equipment and well-planned emergency response procedures will minimize exposure from normal use as well as emergency spill situations.

Exposure

Agricultural workers are often exposed to herbicides when they unintentionally re-enter a treated area too soon following treatment. People who mix and apply herbicides are at the greatest risk of exposure. The most common routes of exposure are through the skin (dermal) or by inhalation (to the lungs). Accidental spills or splashing into the eyes is also possible and with some compounds, can result in severe eye damage and even blindness.

Agricultural herbicide applicators are typically exposed to herbicide levels ranging from micrograms to milligrams per cubic meter of air through inhalation, but exposures through the skin are thought to be much greater (Spear ,1991). Spilling concentrated herbicide on exposed skin can be the toxic equivalent of working all day in a treated field (Libich et al., 1984). Dermal exposure can occur to the hands (directly or through permeable gloves), splashes onto clothing or exposed skin, and anywhere you wipe your hands (e.g., thighs, brow). Some tests have found relatively high levels of dermal exposure to the crotch and seat of workers who got herbicide on their hands, and then touched or wiped the seat of their vehicles. Because adsorption through the skin is the most common route of exposure for applicators (Marer, 1988), the dermal LD_{50} may provide more practical information on the relative toxicity of an herbicide rather than the oral LD_{50}, which is based on oral ingestion.

Toxic Effects

A person's reaction to pesticide poisoning depends on the toxicity of the pesticide, the size of the dose, duration of exposure, route of absorption, and the efficiency with which the poison is metabolized and excreted by the person's body (Marer 1988, Ware, 1991). Different individuals can have different reactions to the same dose of herbicide. Smaller people are, in general, more sensitive to a given dose than are larger people (Marer, 1988).

Herbicides can poison the body by blocking biochemical processes or dissolving or disrupting cell membranes. Small doses may produce no response while large doses can cause severe illness or death. The effects may be localized, such as irritation to the eyes, nose, or throat, or generalized, such as occurs when the compound is distributed through the body via the blood stream. Symptoms can occur immediately after exposure or develop gradually. Injuries are usually reversible, but in extreme cases can be permanently debilitating (Marer, 1988).

Common symptoms of low-level exposure (such as occurs when mixing or applying herbicides in water) to many herbicides include skin and eye irritation, headache, and nausea. Higher doses (which can occur when handling herbicide concentrates) can cause blurred vision, dizziness, heavy sweating, weakness, stomach pain, vomiting, diarrhea, extreme thirst, and blistered skin, as well as behavioral alterations such as apprehension, restlessness, and anxiety (Marer , 1988). Extreme cases may result in convulsions, unconsciousness, paralysis, and death.

Impurities produced during the manufacturing process and adjuvants added to the formulation may be more toxic than the herbicide compound itself. Consequently, LD_{50} is determined for the technical grade of the herbicide may not be the same as that for the brand name formulation. Combinations of herbicides furthermore, can have additive and synergistic effects in which a formulation of two or more herbicides is two to 100 times as toxic as any one of the herbicides alone (Thompson, 1996). Labels should be read carefully for manufacturer's warnings and safety precautions that may be required for a particular formulation.

REFERENCES

1. Bovey, R. W., C. Richardson, E. Burnett, M. G. Merkle, and R. E. Meyer. 1978. Loss of spray and pelleted picloram in surface runoff water. J. Environ. Qual. 7(2):178180.
2. DiTomaso, J. M. 1999. Growth regulators. Presented at the UC Davis Weed Science School, Feb. 24-26 and March 1-3, 1999. Davis, California. EXTOXNET. 1993. Manifestations of toxic effects. Extension Toxicology Network. Toxicology Information Briefs. http://ace.orst.edu/info/extoxnet/tibs/manifest.htm
3. Farmer, W. J., and Y. Aochi. 1987. Chemical conversion of pesticides in the soil-water environment. Chapter 7 in Fate of

pesticides in the environment. J. W. Biggar and J. N. Seiber, eds. University of California, Publication 3320.

4. Freed, V. H., and C. T. Chiou. 1981. Physicochemical factors in routes and rates of human exposure to chemicals. Chapter 3 in Environmental health chemistry. J. D. McKinney, ed. Ann Arbor Science Publishers Inc., Ann Arbor, Michigan.
5. Helling, C. S., P. C. Kearney, and M. Alexander. 1971. Behavior of pesticides in soil. Adv. Agron. 23:147-240.
6. Hutzinger, O. 1981. Environmental and toxicological chemistry at the University of Amsterdam: Five years of philosophy and practice of environmental health chemistry. Chapter 2 in Environmental health chemistry. J. D. McKinney, ed. Ann Arbor Science Publishers Inc., Ann Arbor, Michigan.
7. Kearny, P. C., and J. S. Karns. 1987. Microbial Metabolism. Chapter 10 in Fate of pesticides in the environment. J. W. Biggar and J. N. Seiber, eds. University of California, Publication 3320.
8. Marer, P. J., M. L. Flint, and M. W. Stimmann. 1988. The safe and effective use of pesticides. University of California, Division of Agriculture and Natural Resources. Publication 3324. 387 pps.
9. McCall, P. J., S. A. Vrona, and S. S. Kelley. 1981. Fate of uniformly carbon-14 ring labeled 2,4,5-Trichlorophenoxyacetic acid and 2,4-Dichlorophenoxyacetic acid. J. Agric. Food Chem. 29:100-107.
10. Parker, L. W., and K. G. Doxtader. 1983. Kinetics of the microbial degradation of 2,4-D in soil: effects of temperature and moisture. J. Environ. Qual. 12(4):553-558. Que Hee, S. S., and R. G. Sutherland. The phenoxyalkanoic herbicides. Vol 1, CRC series in pesticide chemistry. CRC Press,
11. Boca Raton, Fla. Spear, R. 1991. Recognized and possible exposure to pesticides. Chapter 6 in Handbook of pesticide toxicology, Vol. 1, General Principles.
12. W.J. Hayes, Jr. and E. R. Laws, Jr. eds. Academic Press, Inc. San Diego, California. 1576 pages.
13. Taylor, A. W., and D. E. Glotfelty. 1988. Evaporation from soils and crops. Chapter 4 in Environmental chemistry of herbicides, Vol I. R. Grover, ed. CRC Press, Boca Raton, Fla.

14. Thompson, H. M. 1996. Interactions between pesticides; a review of reported effects and their implications for wildlife risk assessment. Ecotoxicology 5:59-81.

15. Voos, G. and P. M. Groffman. 1997. Relationships between microbial biomass and dissipation of 2,4-D and dicamba in soil. Biol. Fertil. Soils 24:106-110.

16. Ware, G. W.. 1991. Fundamentals of pesticides: A self-instruction guide. Thomson Publications. Fresno, California. 307 pgs.

17. World Health Organization. 1984. 2,4-Dichlorophenoxyacetic acid (2,4-D), Environmental Health Criteria 29.

18. United Nations Environment Programme, Geneva. 151 pgs. WSSA. 1994. Herbicide Handbook. Weed Science Society of America. Champaign, Illinois. 352pp.

12

CHAPTER

INTRODUCTION TO FUNGICIDES TOXICOLOGY

INTRODUCTION AND TOXICOLOGY OF FUNGICIDES

Fungicides are either chemicals or biological agents that inhibit the growth of fungi or fungal spores. Modern fungicides do not kill fungi, they simply inhibit growth for a period of days or weeks. Fungi can cause serious damage in agriculture, resulting in critical losses of yield, quality and profit. Fungicides are used both in agriculture and to fight fungal infections in animals. Chemicals used to control oomycetes, which are not fungi, are also referred to as fungicides as oomycetes use the same mechanisms as fungi to infect plants (Latijnhouwers et al., 2000). Fungicides can either be contact, translaminar or systemic. Contact fungicides are not taken up into the plant tissue, & only protect the plant where the spray is deposited; translaminar fungicides redistribute the fungicide from the upper, sprayed leaf surface to the lower, unsprayed surface; systemic fungicides are taken up & redistributed through the xylem vessels to the upper parts of the plant. New leaf growth is protected for a short period. Most fungicides that can be bought retail are sold in a liquid form. The most common active ingredient is sulfur, present at 0.08% in weaker concentrates, and as high as 0.5% for more potent fungicides. Fungicides in powdered form are usually around 90% sulfur and are very toxic. Other active ingredients in fungicides include neem oil, rosemary oil, jojoba oil, and the bacterium Bacillus subtilis. Fungicide residues have been found on food for human consumption, mostly from postharvest treatments (Brooks & Roberts, 1999). Some fungicides are dangerous to human health, such as vinclozolin, which has now been removed from use (Hrelia, 1996), FCX and DFB that are used as pesticides to control pests and they have many side effects on natural non-target organisms (Rouabhi et al., 2009). In this chapter, we will develop the fungicides and their toxicity on biological and ecological systems.

CLASSIFICATION OF FUNGICIDES

Different authors have differing classification systems according to chemical structure, which somewhat complicates and confuses both the presentation and the discussion of fungicides. Several classification systems based on structure appear more of a web organization than a rationalized listing. In addition to classification by chemical structural grouping, fungicides can be categorized agriculturally and horticulturally according to themode of application (use). According to the origin of fungicides, we can classify them in two major groups of fungicides

1. ***Biologically based fungicides (biofungicides)***: Contain living microorganisms (bacteria, fungi) that are antagonistic to the pathogens that cause turf disease. Examples: Ecoguard contains Bacillus licheniformis; Bio-Trek 22G contains Trichoderma harzianum. In the case of a biofungicide, the Latin name of the microbe that it contains is the generic name of the fungicide.
2. ***Chemically based fungicides***: Synthesized from organic and inorganic chemicals, most of the fungicides that are sold throughout the world are chemically-based. They can be recognized according to similarities in three groups. There are 29 generic names (active ingredients) associated with turf grass fungicides, shown in the table 12.1.

Table 12. 1: Fungicides generic names according to their chemical structure (Burpee, 2006).

1.	propiconazole	2.	triadimefon
3.	myclobutanil	4.	fenarimol
5.	triticonazole	6.	tetraconazole
7.	fluoxastrobin	8.	trifloxystrobin
9.	azoxystrobin	10.	pyraclostrobin
11.	flutolanil	12.	boscalid
13.	polyoxin D	14.	thiophanate-methyl
15.	iprodione	16.	vinclozolin
17.	mefenoxam	18.	propamocarb
19.	fosetyl aluminum	20.	phosphonate
21.	quintozene	22.	chloroneb
23.	ethazole	24.	mancozeb
25.	thiram	26.	hydrogen dioxide
27.	chlorothalonil	28.	fludioxonil
29.	cyazofamid	30.	Biofungicides

These 29 names represent 16 groups that have similar chemical structures (table 2). It is important to know which fungicides are chemically related to one another. For example, you should know that azoxystrobin, trifloxistrobin, and pyraclostrobin are chemically related to each other. However, they differ chemically from a fungicide, such as propiconazole, which is in a different chemical group. Because (i) all fungicides in a chemical group generally, control the same diseases. For example, the strobilurin fungicides provide good to excellent control of anthracnose, brown patch, gray leaf spot and summer patch. If you have purchased one strobilurin fungicide for control of these diseases, it is probably not necessary to purchase another. (ii) Since all fungicides in a chemical group control the same diseases, it does not make sense to tank-mix fungicides that represent a common chemical group in order to expand the scope of control. For example, to control anthracnose and dollar spot, tank-mixing two strobilurin fungicides will not work well because the strobilurins provide poor to weak control of dollar spot. (iii) If a pathogen develops resistance to one fungicide in a chemical group, the pathogen is usually resistant to all fungicides in that particular group. In Georgia,the fungi that cause dollar spot, Pythium blight and anthracnose have developed resistance to fungicides in one or more chemical groups.

Table 12. 2: Chemical groups of fungicides according the Generic names (Burpee, 2006).

Generic Names	Chemical Group	
propiconazole **triadimefon** **myclobutanil** **triticonazole** **tetraconazole**	triazoles	DMI (demethylation Inhibitors fungicides)
fenarimol	pyrimidines	
fluoxastrobin **trifloxystrobin** **azoxystrobin** **pyraclostrobin**	strobilurins	
polyoxin D	polyoxins	
thiophanate-methyl	benzimidazoles	
iprodione **vinclozolin**	dicarboxamides	
mefenoxam	phenylamides	
propamocarb	carbamates	
fosetyl aluminum **phosphonate**	phosphonates	
mancozeb **thiram**	dithiocarbamates	
quintozene **chloroneb** **ethazole**	aromatic hydrocarbons	
hydrogen dioxide	peroxides	
chlorothalonil	nitriles	
fludioxonil	phenylpyrolles	
cyanofamid	cyanoimidazole	
flutolanil **boscalid**	carboxamides	
Ecoguard **Sonata** **Soilguard**	Biofungicides	

Topical activity Fungicides can be placed into one of four groups based topical activity:

- **Contact fungicides:** Contact fungicides act only on plant surfaces. They are not absorbed by leaves, stems or roots and cannot inhibit fungal development inside plants. Example: dithiocarbamates, nitriles, aromatic hydrocarbons, peroxides, phenylpyrolles, cyanoimidazoles.

- **Localized penetrants:** Localized penetrant fungicides are absorbed by leaves and move short distances within a treated leaf, they do not move from one leaf to another and they are not absorbed by roots. These fungicides inhibit fungi on treated plant surfaces and inside treated leaves. Example: dicarboximides, strobilurins (except azoxystrobin and fluoxastrobin)
- **Acropetal penetrants:** Acropetal penetrants can penetrate plants through roots, shoots and leaves. These fungicides are absorbed by the xylem and move upward (acropetally) in plants. Acropetal penetrants inhibit fungi on and in treated plant surfaces and inside plant parts that lie above the treated surface. Example: benzimidazoles, triazoles, pyrimidines, carboximides, acylalanines, plus the strobilurins azoxystrobin and fluoxastrobin.
- **Systemic fungicides:** Systemic fungicides are the only fungicides that are absorbed into xylem and phloem and moves up and down in plants. These fungicides inhibit fungi on and in treated plant surfaces and inside plant parts that lie above or below the treated surfaces. Example: phosphonates.

Mode of Action

The body or thallus of most fungi exists as microscopic tubes called hyphae (Fig. 1)

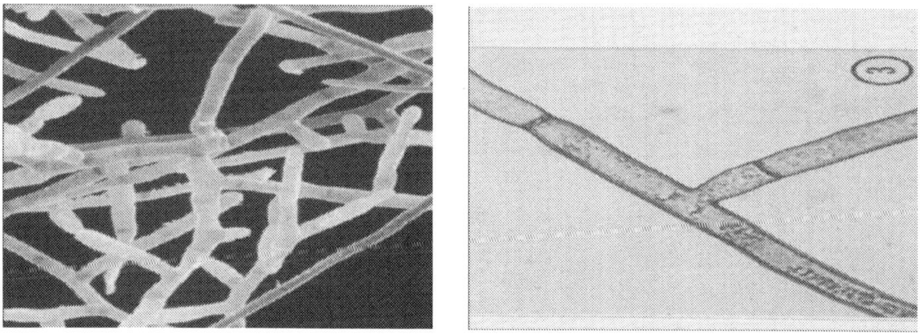

Figure 1: Hyphae of a fungi (Burpee, 2006) A fungal cell contains many of the same organelles as other eukaryotes (Fig.2).

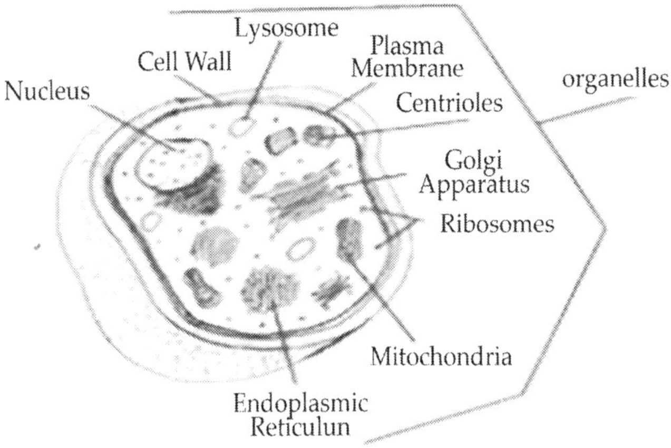

Figure 2: Fungal cell with organelles (Foster and Smith, 2010)

Fungicides can be divided into 2 groups based on mode of action in fungal cells:

a. Site-specific inhibitors: Site-specific inhibitors target individual sites within the fungal cell (Fig. 3).

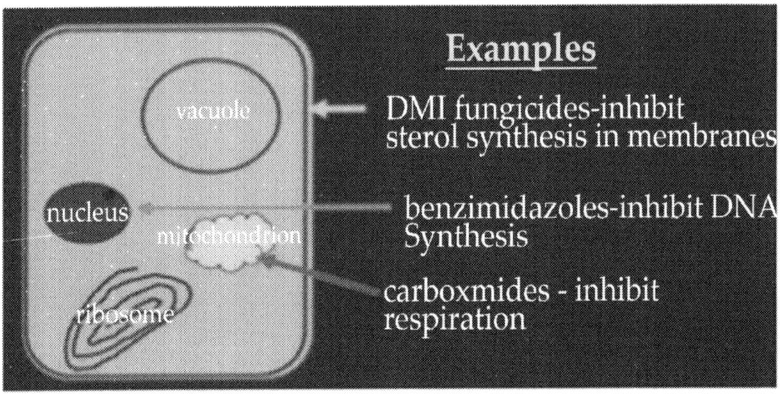

Figure 3: Site-Specific Inhibitors. DMI: demethylation inhibitors fungicides (Burpee, 2006).

b. Multi-site inhibitors Multisite inhibitors target many different sites in each fungal cell (fig. 4)

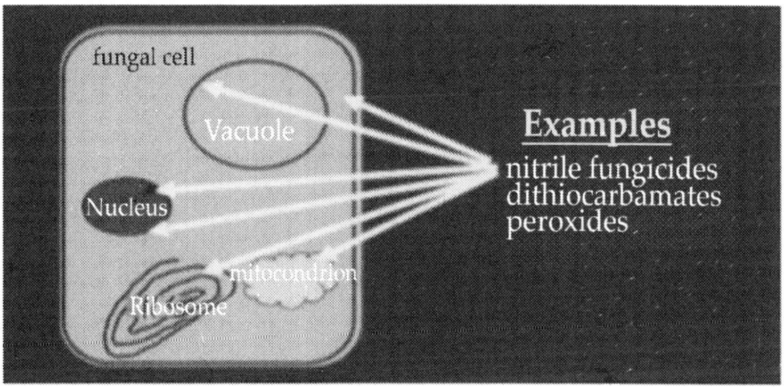

Figure 4: Multi-site Inhibitors (Burpee, 2006).

TOXICOLOGY OF FUNGICIDES

In general, fungicides are of low to moderate mammalian toxicology, although they are believed to have a higher overall incidence than other pesticides to cause developmental toxicology and oncogenesis (Costa, 1997). It has, for example, been estimated that more than 80 per cent of all oncogenic risk from the use of pesticides comes from a few fungicides (NAS, 1987). However, fungicides usually are responsible for only a small proportion of pesticide-related deaths, and account for only about 5 per cent or less of human pesticide exposures reported to Poison Control Centers (Blondell, 1997; Hayes and Vaughn, 1977; Litovitz et al., 1994). It has been noted that since fungi differ significantly in morphology and physiology from other forms of life, they may be successfully combated by compounds of low toxicity to other organisms, notably mammals (Edwards et al., 1991). However, since the mechanism of injury to pathogenic fungi may be different to that for injury to mammalian systems, it is possible that the two properties may co-exist in a given fungicide molecule (Marrs and Ballantyne, 2004).

It has been noted (Phillips, 2001) that the ideal fungicide should have the following characteristics: (a) low mammalian toxicity, (b) low ecotoxicity, (c) low phytotoxicity, (d) high penetration rates for spores and mycelia, and (e) limited biodegradation on the plant surface. Many fungicides combine several of these characteristics but few approach optimum for all of them.

Triazoles

This chemical family of fungicides, introduced in the 1980s, consists of numerous members: difenoconazole, fenbuconazole, myclobutanil, propiconazole, tebuconazole, tetraconazole, triadimefon, and triticonazole. They are important tools against diseases of turfgrasses, vegetables, citrus, field crops and ornamental plants. Homeowner products are available for use as well, and may be readily obtained at garden and nursery retail centers. They are applied as foliar sprays and seed treatments, but are diverse in use, as they may be applied as protectant or curative treatments. If applied as a curative treatment, triazole applications must be made early in the fungal infection process. Once the fungus begins to produce spores on an infected plant, the triazoles are not effective. Although the triazoles do not have the degree of systemic movement of many herbicides, they are xylem-mobile. They are readily taken up by leaves and move within the leaf. The triazoles are very specific in their mode of action – they inhibit the biosynthesis of sterol, a critical component for the integrity of fungal cell membranes. Because their site of action is very specific, there are resistance concerns (Fishel, 2005).

Toxicology

By the oral route of exposure, these triazoles would be considered as having low toxicity. Inhalation of dusts can cause irritation of the nose, lungs, and throat. For mycobutanil, in animals, effects were reported on the following organs: testes, adrenal gland, kidney, and thyroid. Mycobutanil did not cause cancer or birth defects; only doses that caused significant toxicity to parent animals caused reproductive effects on laboratory animals. Increased incidence of liver tumors at extremely high doses was reported in laboratory studies involving male mice who had been exposed to propiconazole or tebuconazole. There were no reproductive, developmental or chronic effects reported with either propiconazole or tebuconazole. Additionally, tebuconazole is considered to not cause any mutagenic orgenotoxic effects; however, EPA has classified it as a "possible human carcinogen" because of the liver effects seen with mice.

The main concern with triadimefon is its potential to cause birth defects, although data suggest that in humans such effects would occur only at moderate to high doses of exposure. Ecologically, the main concern with the triazoles is with fish and other aquatic organisms. Their labels will carry statements expressing this concern in the Environmental Hazards section. Of this pesticide family, only difenoconazole is considered to be highly toxic to fish. Most of the triazoles are considered to be practically nontoxic to birds and bees. Mammalian toxicities for the triazole fungicides are shown in Table 3. Table 4 lists the toxicities to wildlife by the common name of the pesticide (Fishel, 2005).

Table 3: Triazole fungicide mammalian toxicities (mg/kg of body weight).

Common name	Rat oral LD 50	Rabbit dermal LD50
Difenoconazole	1,453	2,010
Fenbuconazole	>2,000	>5,000 (rat)
Mycobutanil	1,600	>5,000 (rat)
Propiconazole	1,517	>4,000
Tebuconazole	1,700	>2,000
Triadimefon	569	2,000
Triticonazole	>2,000	>2,000

Table 4: Triazole fungicide wildlife toxicity ranges.

Common name	Bird acute oral LD 50 (mg/kg)*	Fish (ppm)**	Bee***
Difenoconazole	PNT	HT	PNT
Fenbuconazole	ST	PNT	PNT
Mycobutanil	PNT	MT	PNT
Propiconazole	PNT	MT	PNT
Tebuconazole	PNT	MT	PNT
Triadimefon	PNT - ST	ST	PNT
Triticonazole	PNT	ST	PNT

- *Bird LD 50: Practically nontoxic (PNT) = >2,000; slightly toxic (ST) = 501 – 2,000; moderately toxic (MT) = 51 – 500; highly toxic (HT) = 10 – 50; very highly toxic (VHT) =<10.
- **Fish LC 50: PNT = >100; ST = 10 – 100; MT = 1 – 10; HT = 0.1–1 ; VHT = <0.1.

- ***Bee: HT = highly toxic (kills upon contact as well as residues); MT = moderately toxic (kills if applied over bees); PNT = relatively nontoxic (relatively few precautions necessary).

TOXICOLOGY CYPROCONAZOLE FAMILY

The primary dissipation routes of Cyproconazole (Fig. 5) in surface soil are microbial degradation and plant uptake. Soil photolysis (breakdown by sunlight) and volatilization are not significant routes of degradation. The major breakdown product (degradate, metabolite) of cyproconazole is further broken down to intermediate metabolites, carbon dioxide and bound material. There has been no evidence that Cyproconazole accumulates in the soil.

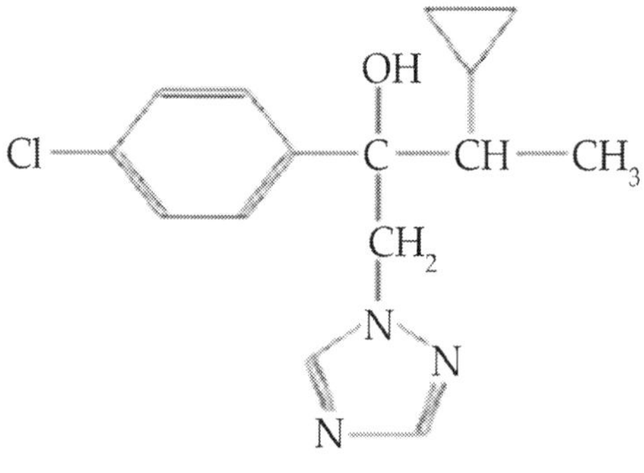

Figure 5: Cyproconazole

Toxicology

Cyproconazole is of low risk to birds, mammals, bees and other non-target terrestrial organisms. Cyproconazole is moderately to slightly toxic to most aquatic organisms, and because of the low use rates the exposure potential is low and hence poses minimal risk of adverse effects (Envirofacts, 2005).

Oral LD50s are 1020mg/kg for male rats, 1333 mg/kg for female rats, 200 mg for male mice, and 218 mg/kg for female mice. The percutaneous rabbit LD50 is >2000 mg/kg. The rat 4-h inhalation is LC50 > 5.65 mg/L. The major plant residue is Cyproconazole. There is moderately rapid soil degradation; DT50 is about 3 months. Avian acute oral LD50 for Japanese quail is 150 mg/kg. Eight-day dietary LC50s are 816mg/kg (diet) for Japanese quail and 1197 mg/kg (diet) for mallard duck. Aquatic organism 96-h LC50 toxicity values include 18.9 mg=L for carp, 19 mg/L for trout, and 21 mg/L for bluegill sunfish. In Daphnia, the 48-h LC50 is 26 mg/L. For bees, the contact LD50 is >0.1 mg/bee and the peroral LD50 is >1mg/bee (Marrs and Ballantyne, 2004).

Plants absorb cyproconazole and rapidly degrade it to multiple metabolites. Studies have shown that cyproconazole residues may exist at harvest, but the levels are insignificant and well under the safety margins for human and environmental risks as established by regulatory authorities in many countries, including the US EPA (Envirofacts, 2005).

Aromatics hydrocarbons fungicides

Major fungicides in this group include chlorothalonil, tecnazine, chloroneb, dichloran, hexachlorobenzene, quintozene, pentachlorophenol, and sodium pentachlorophenate. Many of these are, or metabolized to, uncouplers of oxidative phosphorylation. This can lead to excessive heat production, hyperpyrexia, liver damage, and corneal opacities.

TOXICOLOGY OF CHLORONEB FAMILY

Chloroneb (Fig. 6) is a broad spectrum systemic fungicide taken up by the roots, and used on various fruit and vegetable crops as a wet or dry application powder or dust. Mechanism of fungal toxicity may be related to inhibition of DNA polymerization (Phillips, 2001). Principal use in Soil systemic and supplemental seed treatment for seedling diseases of beans, sugar beets, turf and soybeans. Excellent against damping-off (Harding, 1979-80). Used for the treatment of turf grass to control snow mold (Typhula) and Pythium blight (Meister et al., 1994).

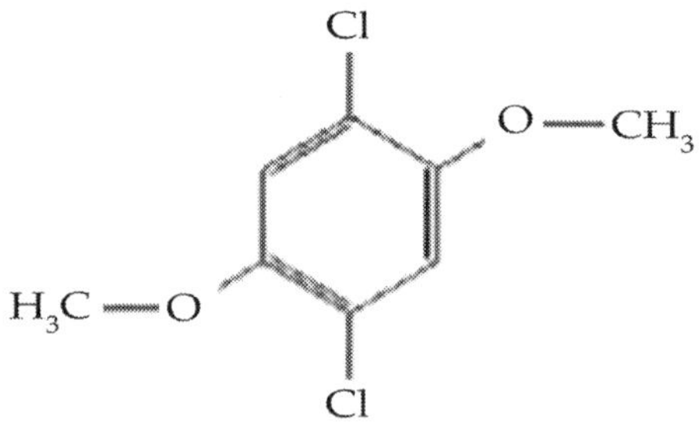

Figure 6: Chloroneb Toxicology

a. Acute Toxicity

Dermal: LD50 = >5000 mg/kg (rabbit). A 50% aqueous suspension of the 65% caused no irritation to guinea-pigs and repeated applications did not result in skin sensitization (Worthing, 1979).

Oral: LD50 = >11,000 mg/kg (rat) (Wothing, 1979)

b. Environmental considerations Hazardous to fish and wildlife. Nonphytotoxic when used as directed. Does not leach from the soil (Harding, 1979-80). The material was not toxic to bluegill Sunfish at 4,200 ppm in the 48-hour exposure.

Dithiocarbamate fungicides

The dithiocarbamate fungicides: ferbam, mancozeb, maneb, nabam, thiram, zineb and ziram were evaluated at the Joint FAO/WHO Meeting in 1967. Although the biochemical data were limited, temporary acceptable daily intakes (ADI's) were established for all of these compounds, but it was pointed out that these ADI's are to be applicable to the parent compounds only (FAO/WHO, 1968).

TOXICOLOGY OF THIRAM FAMILY

Thiram (fig. 7) is a dimethyl dithiocarbamate compound used as a fungicide to prevent crop damage in the field and to protect harvested crops from deterioration in storage or transport. Thiram is also used as a seed protectant and to protect fruit, vegetable, ornamental, and turf crops from a variety of fungal diseases. In addition, it is used as an animal repellent to protect fruit trees and ornamentals from damage by rabbits, rodents, and deer. Thiram is available as dust, flowable, wettable powder, water dispersible granules, and water suspension formulations, and in mixtures with other fungicides. Thiram has been used in the treatment of human scabies, as a sunscreen, and as a bactericide applied directly to the skin or incorporated into soap.

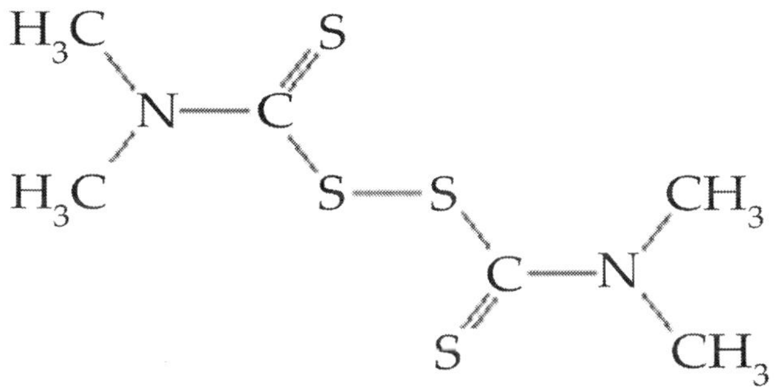

Figure 7: Thiram.

Toxicology

a. Acute Toxicity

Thiram is slightly toxic by ingestion and inhalation, but it is moderately toxic by dermal absorption. Acute exposure in humans may cause headaches, dizziness, fatigue, nausea, diarrhea, and other gastrointestinal complaints. In rats and mice, large doses of thiram produced muscle incoordination, hyperactivity followed by inactivity, loss of muscular tone, labored breathing, and convulsions. Most animals died within 2 to 7 days. Thiram is irritating to the eyes, skin, and respiratory tract. It is a skin sensitizer. Symptoms of acute inhalation exposure to thiram include itching, scratchy throat, hoarseness, sneezing, coughing, inflammation of the nose or throat, bronchitis, dizziness, headache, fatigue, nausea, diarrhea, and other gastrointestinal complaints.

Persons with chronic respiratory or skin disease are at increased risk from exposure to thiram (U.S. National Library of Medicine, 1995). Ingestion of thiram and alcohol together may cause stomach pains, nausea, vomiting, headache, slight fever, and possible dermatitis.

Workers exposed to thiram during application or mixing operations within 24 hours of moderate alcohol consumption have been hospitalized with symptoms. The 4-hour inhalation LC50 for thiram is greater than 500 mg/L in rats. Reported oral LD50 values for thiram are 620 to over 1900 mg/kg in rats; 1500 to 2000 mg/kg in mice; and 210 mg/kg in rabbits (Edwards et al., 1991; Kidd and James, 1991). The dermal LD50 is greater than 1000 mg/kg in rabbits (U.S. National Library of Medicine, 1995) and in rats (Edwards et al., 1991; Kidd and James, 1991).

b. Chronic toxicity

Symptoms of chronic exposure to thiram in humans include drowsiness, confusion, loss of sex drive, incoordination, slurred speech, and weakness, in addition to those due to acute exposure. Repeated or prolonged exposure to thiram can also cause allergic reactions such as dermatitis, watery eyes, sensitivity to light, and conjunctivitis (Edwards et al., 1991). Except for the occurrence of allergic reactions, harmful chronic effects from thiram have been observed in test animals only at very high doses. In one study, a dietary dose of 125 mg/kg/day thiram was fatal to all rats within 17 weeks. Oral doses of about 49 mg/kg/day to rats for 2 years produced weakness, muscle incoordination, and paralysis of the hind legs. Rats fed 52 to 67 mg/kg/day for 80 weeks exhibited hair loss, and paralysis and atrophy of the hind legs. Symptoms of muscle incoordination and paralysis from thiram poisoning have been shown to be associated with degeneration of nerves in the lower lumbar and pelvic regions. Day-old white leghorn chicks fed 30 and 60 ppm for 6 weeks exhibited bone malformations. At doses of about 10% of the LD50 for 15 days, thiram reduced blood platelet and white blood cell counts, suppressed blood formation, and slowed blood coagulation in rabbits (Edwards et al., 1991).

c.Organ toxicity

Studies have shown evidence of damage to the liver by thiram in the form of decreased liver enzyme activity and increased liver weight (Edwards et al., 1991). Thiram may also cause damage to the nervous system, blood, and kidneys (U.S. National Library of Medicine, 1995). d. Ecological Effects on birds: Thiram is practically nontoxic to birds. The reported dietary LC50 of thiram in Japanese quail is greater than 5000 ppm (Hill and Camardese, 1986). Reported dietary LC50 values in pheasants and mallard ducks are 2800 ppm and 673 ppm, respectively (Hudson et al., 1984). The LD50 for the compound in red-winged blackbirds is greater than 100 mg/kg (Kidd and James, 1991).

d. *Ecological effects*

Effects on birds: Thiram is practically nontoxic to birds. The reported dietary LC50 of thiram in Japanese quail is greater than 5000 ppm (Hill and Camardese, 1986). Reported dietary LC50 values in pheasants and mallard ducks are 2800 ppm and 673 ppm, respectively (Hudson et al., 1984). The LD50 for the compound in red-winged blackbirds is greater than 100 mg/kg (Kidd and James, 1991).

Effects on aquatic organisms: Thiram is highly toxic to fish (U.S. National Library of Medicine, 1995). The LC50 for the compound is 0.23 mg/L in bluegill sunfish, 0.13 mg/L in trout, and 4 mg/L in carp (Mayer and Ellersieck, 1986). Thiram is not expected to bio-concentrate in aquatic organisms (Howard, 1989).

BENZIMIDAZOLES FUNGICIDES

This class is confused since the individual classes are closely related. Sometimes, however, the benzimidazoles are classified separately but in an overlapping manner, creating confusion. They are nitrogen heterocyclic compounds, with parent structures of thiabendazole and/or benzimidazole. Included in this overall group are benomyl, thiabendazole, thiophanate, thiophanate–methyl, mebendazole, carbedazim, imazalil, and fuberidazole. Benomyl, carbendazim, thiophanate, and thiophanate–methyl are sometimes referred to (and classified) as benzimidazoles carbamates. Many of these fungicides inhibit mitochondrial fumarate reductase, reduce glucose transport, and uncouple oxidative phosphorylation. Inhibition of microtubule polymerization by binding to -tubulin is a primary action, and specific high affinity binding to host α-tubulin occurs at significantly lower concentrations than mammalian protein binding (Phillips, 2001).

Toxicology of Benomyl Family

Benomyl was first reported as a fungicide in 1968 and introduced onto the UK market in 1971 by the US Company Du Pont (Tomlin, 1994). It is a systemic benzimidazole fungicide that is selectively toxic to microorganisms and to invertebrates, especially earthworms (Extoxnet, 1994). Benomyl and its main metabolite carbendazim bind to microtubules (an essential structure of all cells) and therefore interfere with cell functions such as cell division and intracellular transportation. The selective toxicity of benomyl as a fungicide is possibly due to its heightened effect on fungal rather than mammalian microtubules (WHO/PCS, 1994). Benomyl is used as a pre-harvest systemic fungicide, and as a post-harvest dip or dust. It combats a wide range of fungal diseases of arable and vegetable crops, apples, soft fruit, nuts, ornamentals, mushrooms, lettuce, tomatoes and turf. It is also available widely for amenity and amateur garden use (Whitehead, 1996).

Toxicology

a.Acute Toxicity

Benomyl is of such a low acute toxicity to mammals that it has been impossible or impractical to administer doses large enough to establish an LD50. It therefore has an arbitrary LD50 that is 'greater than 10,000 mg/kg/day for rats'. However, skin irritation may occur with workers exposed to benomyl (Extoxnet, 1994). It is a mild to moderate eye irritant and is a skin sensitizer. Florists, mushroom pickers and flower growers have reported allergic reactions to benomyl (MAFF, 1992). In 1992, benomyl exposure caused adverse occupational health effects (headaches, diarrhea and sexual dysfunction) in agricultural workers in Florida (Agrow, 1992).

b.Chronic toxicity

In a laboratory study, dogs fed benomyl in their diets for three months developed no major toxic effects but did show evidence of altered liver function at the highest dose (150 mg/kg). With longer exposure, more severe liver damage occurred including cirrhosis after two years (Extoxnet, 1994).

c. Carcinogenic effects

The US Environmental Protection Agency classified benomyl as a possible human carcinogen (Office of Pesticide Programs, 1996). There is an element of doubt in this classification because carcinogenic studies have produced conflicting results. A two-year experimental mouse study has shown it probably caused an increase in liver tumors. The Ministry of Agriculture Fisheries and Food (MAFF) takes the view that this was bought about by the hepatoxic effect of benomyl (MAFF, 1992).

d. Reproductive effects

Tests on laboratory animals have shown benomyl can have an effect on reproduction. In one rat study, where the mothers were fed 1,000 mg/kg/day for four months, the offspring showed a decrease in viability and fertility (WHO, 1993). In studies to investigate the effects of benomyl on male reproductive performance, fertility was reduced at all dose levels tested. In another study, a no-effect level of 15mg/kg/day was established based on testicular abnormalities (MAFF, 1992). Permanent reductions occurred in the size of testes and male accessory glands in 100 day old offspring from female laboratory rats receiving 31.2 mg benomyl/kg body weight per day. Rats developed a reduced sperm activity following acute inhalation exposure, acute and sub-chronic oral exposure. The same effect occurred in dogs following a single four-hour inhalation exposure (MAFF, 1992).

e. Environment

Benomyl binds strongly to soil and does not dissolve in water largely. When applied to turf, it has a half-life of three to six months, and when applied to bare soil the half-life is six to 12 months (Extoxnet, 1994).

PIPERAZINES FUNGICIDES

Triforine toxicology

Triforine (fig. 8) is a piperazine derivative used as a systemic fungicide with protectant, eradicant and curative characteristics. It is used for control of powdery mildew, rusts, black rot and scab on cereals, fruit, ornamentals, and vegetables (Royal Society of Chemistry, 1983; Worthing, 1983). Triforine is also active against storage diseases of fruit and suppresses red spider mite activity (Worthing, 1983). Because of its low hazard to beneficial insects, triforine may be used in Integrated Pest Management (IPM) programs. Triforine comes in emulsifiable concentrates, liquid seed treatments, and wettable powder formulations. Triforine is miscible with common insecticides and herbicides in the recommended manner of use (Royal Society of Chemistry, 1983).

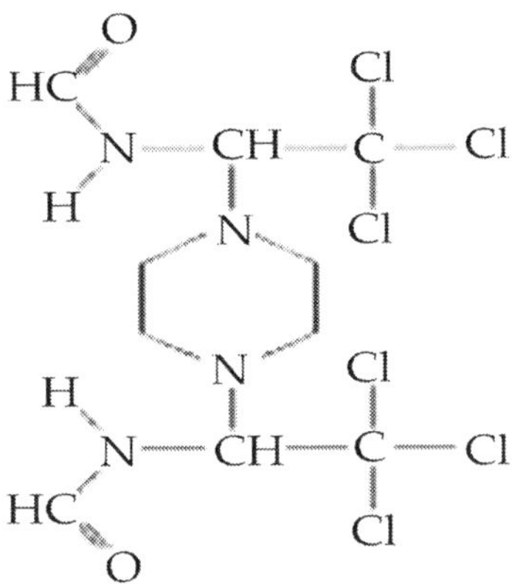

Figure 8: Triforine

Toxicological effects

Acute toxicity

Triforine and the formulated product Saprol have a low acute and dermal toxicity and have a moderate acute inhalation toxicity. The acute oral LD_{50} for triforine in rats is greater than 16,000 mg/kg body weight. The acute percutaneous LD_{50} for rats is greater than 10,000 mg/kg. Acute dermal LD_{50} for rats is greater than 10,000 mg/kg body weight. The one hour acute inhalation LC_{50} for triforine in rats is greater than 4.5 mg/l air (Worthing, 1983). This compound is rapidly absorbed and metabolized by the rat (OHS Database, 1994). The acute oral LD_{50} for the formulated product Saprol in rats is 5,273 mg/kg body weight. The acute dermal LD_{50} for Saprol in rats is 4,186 mg/kg body weight. The acute inhalation LC_{50} for Saprol in rats is greater than 5,288 mg/m3. Saprol is considered an irritant to the skin. The acute oral LD_{50} for triforine in mice is greater than 6,000 mg/kg; and greater than 2,000 mg/kg in dogs. The acute dermal LD_{50} for rabbits is greater than 10,000 mg/kg body weight (Thomson, 1990).

Chronic toxicity

In two-year feeding studies, the No-effect-level (NEL) for triforine in dogs was 100 mg/kg diet and 625 mg/kg diet for rats (Worthing, 1983).

REPRODUCTIVE EFFECTS

A decreased number of fetuses and an increased number of resorptions were observed in a study of pregnant rats fed triforine at a dietary level of 1,600 mg/kg (OHS Database, 1994). The formulated product Saprol does not affect reproduction and development. In another developmental study, rabbits were fed doses of 0, 5, 25 and 125 mg/kg/day of triforine. The maternal No-observable-effect-level (NOEL) was 5 mg/kg/day; the maternal Lowest-effect level (LEL) was 25 mg/kg/day, rabbits exhibited reduced food intake and loss of body weight. The fetotoxic NOEL was 5 mg/kg/day; the fetotoxic LEL was 25 mg/kg/day, decreased average relative weight was observed (U.S Environmental Protection Agency, 1993).

Teratogenic Effects

In a developmental study, rabbits were fed doses of 0, 5, 25 and 125 mg/kg/day of triforine. The teratogenic NOEL was greater than 125 mg/kg/day. The formulated product Saprol is not considered a teratogen (U.S Environmental Protection Agency, 1993).

Mutagenic Effects

The formulated product Saprol is not considered a mutagen.

Carcinogenic Effects

In short and long-term studies of the formulated product Saprol, no irreversible or carcinogenic effects were observed.

Ecological effects

EFFECTS ON BIRDS

The acute oral LD_{50} for triforine in bobwhite quail is greater than 5,000 mg/kg (Worthing, 1983). The formulated product Saprol is practically non-toxic to birds by acute oral exposure and only slightly toxic by dietary exposure. The acute oral LD_{50} for Saprol in bobwhite quail is greater than 5,000 mg/kg. The dietary LC50 for bobwhite quail is 1,850 ppm in the diet. Mallard ducks had a dietary LC_{50} of greater than 4,640 ppm in the diet.

EFFECTS ON AQUATIC ORGANISMS

At 50 mg/l in water, there are no signs of poisoning in Lebistes reticulatus. Rainbow trout and bluegill sunfish tolerate 1,000 mg/l in water for 96 hours without symptoms (Royal Society of Chemistry, 1983). The 96-hour LC_{50} for rainbow trout and bluegill sunfish is greater than 1,000 mg/l (Worthing, 1983). The formulated product Saprol is of low hazard to fish and aquatic invertebrates. Both rainbow trout and bluegill sunfish had a 96-hour LC_{50} of greater than 500 mg/l. The aquatic invertebrate Daphnia (water flea) had a 48-hour EC50 of greater than 25 mg/l. Saprol was also noted to be of low hazard to Scenedesmus subspicatus (aquatic algae). The 96-hour EC_{50} was greater than 380 mg/l.

Effects on Other Animals (Non-target species)

No toxic effect was observed in honeybees at less than or equal to 1,000 mg/kg diet (Worthing, 1983). Triforine and the formulated product Saprol are considered of low hazard to honeybees and to the predatory mite Typhlodromus pyrii. It is also of low hazard to earthworms at recommended dose rates (Meister et al., 1994).

Aliphatic aldehydes Fungicides

Several aliphatic aldehydes are used as fungicides, amongst them formaldehyde and formaldehyde releasers, which have been considered in detail elsewhere (Feinman, 1988).

Toxicology of Acrolein Acrolein reacts with SH groups. It is formulated as a liquid (fig. 9).

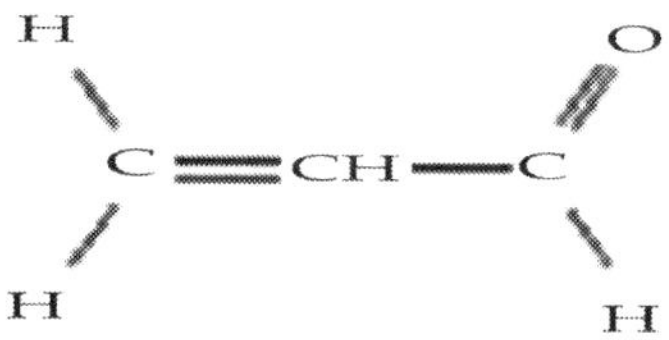

Figure 9: Acrolein.

Toxicology

Acute toxicity

The rat oral LD50 is 29mg=kg and the mouse oral LD50 is 13.9mg=kg (males) and 17.7mg=kg (females). The percutaneous rabbit LD50 is 231mg=kg. By inhalation the rat 1-h LC50 is 65mg=m3 (males) and 60mg=m3 (females): the rat 4-h LC50 is 18.5mg=m3 (males) and 22mg=m3 (females). The rat 30-min LC50 is 131 ppm and the 10-min LC50 is 355 ppm (Ballantyne et al., 1989).

Short-term and subchronic toxicology

Hamsters, rats, and rabbits were exposed to acrolein vapor at 0, 0.4, 1.4, and 4.9 ppm for 6 h/day for 5 days/week for 13 weeks. At 4.9 ppm there was mortality, ocular and nasal irritation, depression of growth, and inflammation, necrosis, hyperplasia, and metaplasia of the respiratory tract epithelium. A no-effect concentration was not established for the rat (Feron et al., 1978).

Chronic toxicology

Sprague-Dawley rats were given acrolein by gavage at daily dosages of 0, 0.05, 0.5, and 2.5mg/kg up to 102 weeks. The only effects noted were decreased serum creatinine kinase and increased early cumulative mortality. There were no significant increases in neoplastic or non-neoplastic histopathology (Parent et al., 1992a).

CARCINOGENICITY

Given by intraperitoneal injection to male Fischer 344 rats, Acrolein had an initiating activity for urinary bladder carcinogenesis (Cohen et al., 1992).

Reproductive toxicology

Male and female rats were incubated and given 70 daily doses of Acrolein at 0, 1, 3, or 6mk=kg. The F0 generation was assigned to a 21-day period of co-habitation and dosing of females continued through co-habitation, gestation, and lactation. F1 generation pups were similarly treated. In general, reproductive indices were unaffected, with the exception of reduced pup weights in the F1 generation at the high dose. Gastric lesions were consistently found in the high dose and some mid-dose animals; erosions of the glandular mucosa and hyperplasia=hyperkeratosis of the fore stomach were the most frequent lesions. Relative to the controls, mortality and body weight gain decreases were noted for high dosage animals (Parent et al., 1992).

BIOLOGICAL FUNGICIDES (BIOFUNGICIDES)

Biofungicides are microorganisms (microbial pesticides) and naturally occurring substances that control diseases (biochemical pesticides) that are approved for organic production. Biofungicides are widely used by organic vegetable growers to control selected foliar and soil borne diseases of vegetable crops (see Table 5). Biofungicides can be applied as a standalone treatment to control a target disease, provided the application is made before the disease starts (Francis and Keinath, 2010). In the case of a biofungicide, the Latin name of the microbe that it contains is the generic name of the fungicide.

Toxicology of EcoGuard

EcoGuard is a concentrated suspension of spores of Bacillus licheniformis SB 3086 that has been found effective as a natural inhibitor of a variety of agronomically important fungal diseases - particularly dollar spot and anthracnose. EcoGuard allows you to control the disease and significantly improve overall turf quality at the same time. The activity of EcoGuard is due to the synthesis of powerful anti-fungal compounds that inhibit fungal growth. As a primary benefit, EcoGuard can be used just like other fungicides. As a secondary benefit, when EcoGuard is integrated with conventional turf management practices, you will see a noticeable and often dramatic improvement in the health and vigor of your turf. In addition, the turf will also recover more quickly from diseases with improved color and increased density in the damaged areas (Novozymes Biologicals Inc., 2007).

B. licheniformis SB3086 is a naturally occurring, ubiquitous bacterium originally isolated from United States farm soil. Consequently, the United States Environmental Protection Agency (USEPA) required limited data for federal registration of EcoGuardTM Biofungicide. The data from acute toxicity/pathogenicity studies on the active ingredient indicate that Bacillus licheniformis SB3086 is not toxic, infective or pathogenic via the oral or inhalation routes of exposure (tested at 1 x 108 Colony Forming Units (CFU) per animal), or via intravenous injection (tested at 1x10' CFU/animal). The end product was not very acutely toxic via oral, dermal, or inhalation routes of exposure. It was also not irritating to the eyes (tested on rabbits) or a dermal sensitizer (tested on guinea pigs), but was a slight dermal irritant (tested on rabbits)

The USEPA waived the requirement for subchronic, chronic, developmental, reproductive toxicity, genotoxicity and oncogenicity studies for federal registration of EcoGuardTM Biofungicide. Instead, the USEPA used reports from the scientific literature to evaluate this product. The data from these reports suggest that B. licheniformis is occasionally associated with infections in individuals who have significant preexisting health problems such as severe immune system depression, cancer or trauma. In addition, it has been associated with reproductive failures (spontaneous abortions and inflammation of the placenta) in cattle, sheep and swine, usually in association with the ingestion of moldy hay.

In addition, it has been associated with reproductive failures (spontaneous abortions and inflammation of the placenta) in cattle, sheep and swine, usually in association with the ingestion of moldy hay. A search of the toxicological literature did not find any additional significant information on B. licheniformis. A quantitative worker risk assessment was not provided in the registration package. However, B.

licheniformis has been used in various industrial fermentation processes for a number of years and, according to the registrant, no pathogenicity, toxicity or hypersensitivity has been reported among these workers

Table 5: List of bio-fungicides (biological) used to control selected vegetable crop diseases (Francis and Keinath, 2010)

Product	Active Ingredient	Disease	Treatment Site
Ballad	*Bacillus pumilus*	Several (Foliar)	Foliar
Bio-Save	*Pseudomonas syringe*	Post-harvest	Irish and sweet potatoes in storage
Contans	*Coniothyrium minitans*	White Mold	Soil applied
Kodiak	*Bacillus subtilis*	*Pythium, Rhizoctonia, Fusarium*	Seed treatment, beans only
Mycostop	*Streptomyces griseoviridis*	Several	Greenhouse; Soil applied
Regalia	Plant extract	Powdery mildew	Foliar
RootShield Granules, RootShield WP	*Trichoderma harzianum*	*Pythium, Rhizoctonia, Fusarium*	Soil applied
Serenade	*Bacillus subtilis*	Powdery mildew, other foliar diseases	Foliar
T22-HC	*Trichoderma harzianum*	*Pythium, Rhizoctonia, Fusarium*	Soil applied
Surround	Kaolin	Powdery mildew	Foliar
Trilogy	Neem Oil	Powdery mildew	Foliar
Actinovate AG	*Streptomyces lydicus*	*Fusarium, Rhizoctonia, Pythium, Phytophthora*	Soil applied Foliar
SoilGard	*Gliocladium virens*	Damping off	Greenhouse-transplants, soil applied

In addition, it has been associated with reproductive failures (spontaneous abortions and inflammation of the placenta) in cattle, sheep and swine, usually in association with the ingestion of moldy hay. A search of the toxicological literature did not find any additional significant information on B. licheniformis. A quantitative worker risk assessment was not provided in the registration package. However, B. licheniformis has been used in various industrial fermentation processes for a number of years and, according to the registrant, no pathogenicity, toxicity or hypersensitivity has been reported among these workers

Given the use pattern of the EcoGuardTM product, exposure of applicators would likely be less than that of fermentation workers. Also, the product label requires applicators and other handlers to wear a long-sleeved shirt, long pants, and shoes plus socks. In addition, handlers must also use a non-powered air purifying NIOSH approved respirator.

The limited toxicity data required to support the federal registration of EcoGuardTM Biofungicide indicate that this product is not very toxic following acute exposures. The active ingredient B. licheniformis also appears to have a low degree of pathogenicity and infectivity to animals and humans. Although there are no animal study data on longer-term exposure, significant risks to workers or the general public from EcoGuardTM use are not expected given the use pattern and the required personal protective equipment. The required personal protective equipment also should protect against the slight dermal irritation that EcoGuardTM may cause.

ECOLOGICAL RISK

B. licheniform is SB3086 is not toxic, infective, or pathogenic to mammals when administered orally, by inhalation, direct tracheal injection, dermally, and by intravenous injection. A significant decrease in weight gain in young mallard ducks was observed when administered EcoGuardTM formulation at a rate of approximately 1.0 ml/kg body weight. The effect was attributed to formulation ingredients other than the active ingredients. B. licheniformis SB3086 appears not to be toxic, infective, or pathogenic to mallards; there were no mortalities.

The 30-day EcoGuardTM rainbow trout LC50 is greater than 1.1 xl06 CFU/ml which is roughly I I 7X (times) the expected environmental concentration (EEC) when the maximum application rate is applied directly to the surface of six-inch deep-water body. The daphnia No Observable Adverse Effect Concentration (NOAEC) is 120X the EEC. There was no sign of infection or pathogenicity in either study.

The EcoGuardTM formulation had no effect on honeybee larva when exposed to B. licheniformis SB3086 at 1.6 x106 CFU/ml, roughly 2/3 full strength formulation, in their diet. No adverse behavioral or developmental abnormalities were observed in emerged adult honeybees that had been exposed as larva. All marine/estuarine organism and nontarget plant testing was waived by the USEPA.

Bacillus licheniformis is a ubiquitous soil organism. While numbers of B. licheniformis in soil are unknown and likely vary from soil to soil, the total number of Bacillus organisms is estimated to be 10' CFU/g soil. The added soil density of B. licheniformis from the proposed use rates would be 0.42% to 1.5%. This is a very small proportion of the naturally occurring bacilli in soil and is not expected to add substantially to the effects of the normally occurring Bacillus populations. No adverse effects to fish or wildlife resources are expected from use of EcoGuardTM Biofungicide when used as labeled. There are reports in the literature of B. licheniformis being a sporadic mammal pathogen. B. licheniformis diseases appear to be limited to cows, sheep, and swine as very unusual events associated with the ingestion of moldy hay. Wild ruminants' exposure to the combination of conditions seemingly implicated in livestock disease should be minimal. The IBA contained in EcoGuardTM poses no risk to fish or wildlife resources: The IBA concentration in EcoGuardTM is very low, application at the maximum label rate results in an IBA application rate of roughly 30 milligrams/acre (Serafini, 2003). It is to note that this study showed a less or more toxicity of all category of chemical fungicides, contrarily, to biofungicides that showed a little or note side effects on human and ecosystems.

REFERENCES

1. Agrow, (1992). More problems for Benlate? 13 March 1992, 13 p.

Ballantyne, B.; Dodd, D.E.; Pritts, I.M.;

2. Nachreiner, D.J. & Fowler, E.H. (1989). Acute vapor inhalation toxicity of acrolein and its influence as a trace contaminant in 2-methoxy-3,4-dihydro-2H-pyran. Hum. Toxicol., Vol., 8: 229–235.
3. Blondell, J. (1997). Epidemiology of pesticide poisoning in the United States, with special reference to occupational cases. Occup. Med., State of the Art Rev., Vol., 12:209–220.
4. Brooks, G.T. & Roberts, T.R. (1999). Pesticide Chemistry and Bioscience. Published by the Royal Society of Chemistry. Burpee, L. (2006). Integrated disease management, an introduction to Fungicides. Courses support.
5. Cohen, S.M.; Garland, E.M.; St John, M.; Okamura, T. & Smith, R.A. (1992). Acrolein initiates rat urinary bladder carcinogenesis. Cancer Res.,Vol., 52: 3577–3581.
6. Costa, L.G. (1997). Basic toxicology of pesticides. Occup. Med. State of the Art Rev. Vol., 12:251–268.
7. Edwards, I.R., Ferry, D.G. & Temple, W.A. (1991). Fungicides & related compounds, In: Handbook of Pesticide Toxicology. Hayes, W.J. & Laws, E.R., Eds. Academic Press, New York, NY,. Vol., 3, pp: 1409–1470
8. Envirofacts, (2005). CYPROCONAZOLE The Active Ingredient in Alto® One of the Active Ingredients in Quadris Xtra®. Syngenta Crop Protection Inc., Greensboro, NC 27419-8300.
9. Extoxnet, (1994). Pesticide Management Education Program, Cornell University, NY. FAO/WHO. (1968). Evaluations of some pesticide residues in food.
10. FAO/PL: 1967/M/11/1; WHO/Food Add./68.30. Feinman, S.E. (1988). Formaldehyde Sensitivity and Toxicity. CRC Press, Boca Raton,
11. FL. Feron, V.J.; Kruysse, A.; Til, H.P. & Immel, H.R. (1978). Repeated exposure to Acrolein vapor: sub-acute studies in hamsters, rats and rabbits. Toxicology, Vol., 9: 47–57.
12. Fishel, F.M. (2005). Pesticide Toxicity Profile: Triazole Pesticides. University of Florida, IFAS extension. PI68.
13. Foster, and Smith, (2010). Germs: Viruses, Bacteria, and Fungi.

Veterinary & Aquatic Services Department Foster & Smith, Inc. Wisconsin. USA.

14. Francis, R. and Keinath, A. (2010). Biofungicides and chemicals for managing diseases in organic vegetable production. CLEMSON Cooperative extension. Information leaflet 88.
15. Harding, W.C. (1979-80). Pesticide profiles, part two: fungicides and nematicides. Univ. Maryland, Coop. Ext. Service Bull. 283, 22 pp.
16. Hayes, W.J and Vaughn, W.K. (1977). Mortality from pesticides in the United States from 1973 to 1976. Toxicol. Appl. Pharmacol. Vol., 42: 235–252.
17. Hill, E.F. & Camardese, M.B. (1986). Lethal Dietary Toxicities of Environmental Contaminants to Coturnix, Technical Report Number 2. U.S. Department of Interior, Fish and Wildlife Service, Washington, DC., 4-37
18. Howard, P. H., (1989). Handbook of Environmental Fate and Exposure Data for Organic Chemicals: Pesticides. Lewis Publishers,
19. Chelsea, MI., 4-20 Hrelia, R. (1996). The genetic and non-genetic toxicity of the fungicide. Vinclozolin Mutagenesis. Vol., 11: 445-453.
20. Hudson, R.H.; Tucker, R.K. & Haegele, M.A. (1984). Handbook of Acute Toxicity of Pesticides to Wildlife, Resource Publication 153. U.S. Department of Interior, Fish and Wildlife Service, Washington, DC., 4-15
21. Kidd, H. and James, D.R., (1991). The Agrochemicals Handbook, Third Edition. Royal Society of Chemistry Information Services, Cambridge, UK. Latijnhouwers, M.; de Wit, P.J. and Govers, F. (2000). Oomycetes and fungi: similar weaponry to attack plants. Trends in Microbiology. Vol., 11: 462-469.
22. Litovitz, T.A.; Felberg, L. & Soloway, R.A. (1994). Annual reports of the American Association of Poison Control Centers. Toxic exposure surveillance system.
23. Am. J. Emerg. Med. Vol., 13: 551–597. MAFF, (1992). Benomyl evaluation No. 57, July 1992, pp: 91-111.
24. Marrs, C.T. & Ballantyne, B. (2004). Pesticides toxicology and

international regulation. John Wiley and Sons, Ltd. 554 pages.

25. Mayer, F. L. and Ellersieck, M. R. (1986). Manual of Acute Toxicity: Interpretation and Data Base for 410 Chemicals and 66 Species of Freshwater Animals. Resource Publication 160. U.S. Department of Interior, Fish and Wildlife Service, Washington, DC,. 4-18
26. Meister, R.T.; Berg, G.L.; Sine, C.; Meister, S. & Poplyk, J. (1994). Farm Chemicals Handbook, 70th Eds. Eds. Meister Publishing Co.,
27. Willoughby, OH. NAS, (1987). Regulating Pesticides in Food. The Delaney Paradox. Report of Committee on Scientific and Regulatory Issues. Unlikely Pesticide Use Patterns. National Academy of Sciences, National Academy Press, Washington,
28. DC Novozymes Biologicals, Inc. (2007). Roots EcoGuard biofungicide. www.novozymes.com/roots.
29. Office of Pesticide Programs, (1996). List of Chemicals Evaluated for Carcinogenic Potential, US EPA, Washington,
30. US. OHS Database. 1994. Occupational Health Services, Inc. 1994. MSDS for Vernolate. OHS Inc.,
31. Secaucus, NJ. Parent, R.A.; Caravello, H.E. and Hiberman, A.H. (1992b). Reproduction study of Acrolein on two generations of rats. Fund. Appl. Toxicol., Vol., 29: 228–237.
32. Parent, R.A.; Caravello, H.E. and Long, J.E. (1992a). Two-year toxicology and carcinogenicity study of acrolein in rats. J. Appl. Toxicol., Vol., 12: 131–139.
33. Phillips, S.D. (2001). Fungicides and biocides. In: Clinical Environmental Health and Toxic Exposures, Sullivan, J.B. & Krieger, G.R., Eds. Lippincott Williams and Wilkins, Philadelphia, 2nd Eds. pp:1109–1125.
34. Rouabhi, R.; Djebar, H. & Djebar, M.R. (2009). Toxic Effects of Combined Molecule from Novaluron and Diflubenzuron on Paramecium caudatum.
35. Am-Euras. J. Toxicol. Sci. Vol., 1(2). September 2009. 74-80. ISSN: 2079-2050; EISSN: 2079-2069.
36. Royal Society of Chemistry (1983). The Agrochemicals Handbook. The University, Nottingham, England.
37. Serafini, P.M. (2003). Bacillus licheniformis SB3086 NYS DEC

Letter - New Active Ingredient Registration 8/03. New York State Department of Environmental Conservation.

38. Thomson, W.T. (1990) Agricultural Chemicals, Book IV: Fungicides. Thomson Publications,
39. Fresno, CA. Tomlin, C., (1994). The Pesticide Manual, 10th Edition, British Crop Protection Council/Royal Society of Medicine. U.S.
40. Environmental Protection Agency, (1993). Office of Pesticides. TOX Oneliners Triforine. April, 1993.
41. U.S. National Library of Medicine, (1995). Hazardous Substances Data Bank. Bethesda, MD., 4-5
42. Whitehead, R. (1996). The UK Pesticide Guide, British Crop Protection Council/CAB International.
43. World Health Organization, (1993). Benomyl, Environmental Health Criteria No 148, Geneva, Switzerland.
44. World Health Organization, (1994). Data sheet on benomyl, WHO/PCS/94.87, Geneva.
45. Worthing, C.R. (1979). The Pesticide Manual: A World Compendium, 6th Eds. The British Crop Protection Council, Croydon, England. 655 pp.

13

CHAPTER

ORGANIC AND BOTANICAL PESTICIDES: CHEAPER & EFFECTIVE PEST CONTROL

INTRODUCTION

It is not the intention of this chapter to entirely eliminate the use of beneficial agricultural chemicals that help to suppress and control destructive pests and diseases and provide food nutrients to the plants. The government also encourages the adoption of the latest improved technology and farming systems that are environmentally and ecologically friendly. The government encourages farmers to adopt practical Integrated Pest Management (IPM) and to attempt in helping and teach interested farmers adopt natural and practical farming systems that will reduce the use of costly and toxic agricultural chemicals and replace them with organic preparations the farmers themselves can produce and use. This can result to producing healthful organic food at lower cost with higher productivity as the soil and environment improves. This may be debatable, but possible and attainable. There is a worldwide need for safe, chemical free food and a demand for organically grown food crops. The alarming increase in the use of toxic chemicals to control pests and diseases on both farm animals and plants has endangered the environment and reduced bio diversity as well as the health of consumers. Farmers have become too dependent on pesticides and fertilizers as the soil has become poorer and insect pests have become resistant to chemicals used. The world that we have created is dominated by a disregard for Nature and a greed that destroys for profit the environment, the ecosystem and the capacity of the land to produce without artificial human interventions. Useful insects and predators of pests are killed together with the insect pests with the wide and intensive use of toxic agricultural chemicals. The surviving pest finds no natural enemies; thus, they increase rapidly resulting in more destructive infestation with more resistant pest to chemical control.

People have to recognize the natural laws that govern and balance nature, its biodiversity and life itself. This investigative study can help farmers reduce their production cost and become self-reliant with renewable crop protection products they can grow and formulate in their own farms, together with their commercial crops and livestock. They can also make their own pesticide and soil amendments that will turn their land into productive farms. This can make farmers self-reliant and self-sustaining through Natural Farming, and can feed the increasing population of the world with healthful food, free from toxic chemical residue especially in rice production which is a main staple for many developing countries.

HISTORY OF ORGANIC PESTICIDE

A wide range of substances and mixtures of substances, highly effective in preventing, destroying, repelling or mitigating pests: insects, mice, unwanted plants, fungi, microorganisms such as bacteria and viruses, and prions are classified as pesticides. User benefit from pest control is obvious, in terms of increased agricultural production and elimination of pest-induced health problems. However, there are now evidence that pesticides do create risk to man and his environment. Therefore, the agricultural use of some persistent organochlorine insecticides, namely DDT, was banned after the 1960s. Other synthetic chemicals, organophosphates (1960s), carbamates (1970s) and pyrethroids (1980s) as well the emerging biopesticides (2000s) came on to substitute them.

Biopesticides are used to control pests, pathogens, and weeds by a variety of means. Microbial biopesticides may include a pathogen or parasite that infects the target. Alternatively, they might act as competitors or inducers of plant host resistance. Biochemical biopesticides can also act through a variety of mechanisms. Some act by inhibiting the growth, feeding, development or reproduction of a pest or pathogen. Still other biopesticides may be used to form a barrier on the host, so as to act as a feeding or infection inhibitor.

Plant extracts were likely the earliest agricultural biopesticides, as history records that nicotine was used to control plum beetles as early as the 17th century. Experiments involving biological controls for insect pests in agriculture date back as far as 1835, when Agostine Bassi demonstrated that w hite-muscadine fungus (*Beauveria bassiana*) could be used to cause an infectious disease in silkworm. Experiments with mineral oils as plant protectants were also reported in the 19th century. During the rapid institutional expansion of agricultural research during the early 20th century, an ever-growing number of studies and proposal for biopesticides were developed.

The first, and still most, widely used biopesticide included spores of the bacteria *Bacillus thuringiensis* (Bt). In 1901, Bt was isolated from a diseased silkworm by Japanese biologist Shigetane Ishiwata. Ernst Berliner in Thuringen, Germany, then rediscovered it ten years later in a diseased caterpillar of flour moth. The Bt pathogen was classified in 1911 as type species *Bacillus thuringiensis* and remains the most widely used biopesticide to this day. In the early 1920 s, the French began to use Bt as a biological insecticide. The first commercially available Bt product, Sporeine, appeared in France in 1938. In the US in the 1950 s, widespread use of biopesticides began to take hold as a host of research on Bt efficacy was published.

In the latter half of the 20th century, research and development continued at a low level because of the widespread adoption of cheaper but more toxic synthetic chemical insecticides. During this time, new products were developed and applied; especially in niche markets where petroleum based chemicals were not registered, not effective, or not economical. For example, in 1956, the Pacific Yeast Product Company developed an industrial process known as submerged fermentation, which allowed production of Bt on a large scale. In 1973, *Heliothis* NPV was granted exemption from tolerance and the first viral insecticide, Elcar received a label in 1975. In 1977, *Bacillus thuringiensis* var. *israelensis* (toxic to flies) was discovered, and in 1983 the strain *tenebrion is* (toxic to beetles) was found. In 1979, the U.S. EPA registered the first insect pheromone for use in mass trapping of Japanese beetles. In the 1990s, researchers began testing kaolin clay as an insect repellent in organic fruit orchards. It was made commercially available, particularly for use in organic systems, in 1999.

Biopesticide development for the control of plant diseases has undergone a similar transformation. During the early 20th century, studies of soil microbiology and ecology had led to the identification of many different microorganisms that act as antagonists or hyperparasites of pathogens and insect pests. A number of these were shown to be useful in field-scale inoculations, but few were developed commercially because of the rapid adoption of chemical pesticides during that time period. Commercial success stories from the 1980s and 1990s include products containing*Agrobacterium radiobacter* for the prevention of crown gall on woody crops and *Pseudomonas fluorescen s* for the prevention of fireblight in orchards where the streptomycin had been overused and resistant pathogen populations were abundant. In the greenhouse and potting mix industry, products containing a variety of microbes that suppressed soilborne pathogens were introduced into the market.

As the costs of overusing such synthetic chemicals became apparent, there was resurgence in academic and industrial research related to biopesticide development. And with the rapid expansion of organic agriculture during the past decade, adoption rates have rapidly increased. Because of this, development of new and useful biopesticides has continued to increase rapidly since the mid-1990 s. In fact, more than 100 biopesticide active ingredients have been registered with the U.S. EPA Biopesticides division since 1995. Many of these have been introduced Biopesticides division since 1995. Many of these have been introduced commercially in a variety of products. Many of the active ingredients currently approved for use in the U.S.A. can be found in publicly available databases.

Sample of Biopesticides Develop Recently Used

***Agrobacterium radiobacter* Strain K84:** *Agrobacterium radiobacter* Strain K84 is a naturally occurring bacterium found in many soils and in plant root zones. This biopesticide is used in the greenhouse and nursery environment to control crown gall, an important plant disease.

Bacillus spp. *Bacillus licheniformis, B. pumilus*, and *B. subtilis* are naturally occurring soil bacteria with fungicidal properties that together have become one of the fastest growing biopesticides in today's market. Successes include uses as seed treatments or dressings, foliar application and soil-applied control of diseases in a variety of crops.

Coniothyrium minitans: *Coniothyrium minitans* is a naturally occurring fungus used commercially to control common *Sclerotinia* plant diseases through parasitism of the resting structures of the pathogen.

Paecilomyces fumosoroseus and P. Lilacinus: *Paecilomyces fumosoroseus* is a naturally occurring fungus used in a greenhouse environment to control several species of insects including whiteflies, thrips, aphids, and spider mites. *Paecilomyces lilacinus* is used to control nematodes that attack plant roots in field crops including many vegetables, fruit, turf, and ornamental crops.

***Trichoderma* spp.** *Trichoderma spp.* is another biopesticide technology developed in the 1990 s that has been widely commercialized in recent years. *Trichoderma*is a genus of fungi that helps to control plant disease by stimulating plant host defenses and growth, and, under certain conditions, parasitizing harmful fungi within the plant root zone.

Azadirachtin: *Azadirachtin* is an insect growth regulator derived from neem tree seeds. Known to affect some 200 species of insects, *azadirachtin* disrupts insect feeding and inhibits its ability to molt as it changes from the pupa to adult stage.

Beauveria bassiana: *Beauveria bassiana* is a naturally occurring soil fungus that grows as white mold. This insect pathogen can be used to control a wide range of target pests, which become infected and develop white muscadine disease, killing the pest within a matter of days.

Cydia pomonella : *granulo virus* (CpGV) CpGV is a natural pathogen of the codling moth, a major pest of tree fruits such as apples and pears. Developed through research begun in the 1980's, commercial use of CpGV in both organic and conventional systems has gained in popularity over the last ten years as codling moth has displayed resistance to many traditional insecticides.

Dysphania ambrosioides An extract of the plant *Dysphania ambrosioides (*syn. *Chenopodium ambrosioides)* is used to control a number of sucking insect pests such as aphids, leafhoppers, whiteflies, and mites in citrus, grapes, tree nuts, and vegetables. This product breaks down the pest's exoskeleton, adversely affects its respiratory system, and interrupts its ability to navigate (find food).

HISTORICAL BACKGROUND OF ORGANIC PESTICIDE USE

The white Chrysanthemum flower, pyrethrum, is mentioned in early Chinese history and it is believed to have passed into Europe along the silk routes. However, the species of the plant was unknown, so for convenience the history of pyrethrum usually starts with the mention in 1847 of the identified species Chrysanthemum cinerariaefolium found growing in Dalmatia (now part of Croatia). Later, ground pyrethrum flowers could be found for sale in most European pharmacies as 'Dalmatian Insect Powder'. From 1885, bales of dried flowers were exported to the USA and, during 1913, 500 tonnes were shipped. The outbreak of World War I ended these exports and the USA started importing dried flowers from Japan, which had started growing the crop in 1886. The primary uses for the ground flowers were for the control of body lice on humans and animals, and crawling insects in the home.

In 1917, the US Navy made the first pyrethrum extracts by percolating ground flowers with kerosene, which were then incorporated into space sprays for use against houseflies and mosquitoes. Pyrethrum products became available for general use even though the chemistry of the active ingredient(s) in pyrethrum was unknown. In 1924, when Staudinger and Ruzicka, working in Switzerland, published a series of classical papers on the chemistry of pyrethrum which showed that the insecticidal properties were due to the presence of two esters which they named Pyrethrin I and Pyrethrin II. This work permitted other chemists to develop analytical methods and these were used by F. Tattersfield at Rothamsted Experimental Station at Harpenden, UK, to undertake a fundamental study "Pyrethrum Flowers – A Quantitative Study of their Development". The combination of identification of the active ingredients, followed by their chemical assay linked with a study of the development of the active principles in the flowers (Tattersfield, 1931) set pyrethrum on a technically orientated path where it has prospered as a unique botanical biopesticide, differing from many others whose active molecules could not be easily analyzed.

The annual world production of dried flowers has rarely exceeded 20,000 metric tonnes and, with an average pyrethrins content of 1.5%, the potential yield is 30 kg of 50% extract per tonne. However, with losses at various processing stages the actual yield is nearer 25 kg, giving a potential annual world production of 500 tonnes of 50% pale extract. Annual sales at consumer level, estimated using confidential data from the largest formulators, indicate a figure of $300 million. Industrial usage related to food processing is probably approximately another $100 million, and minor uses are probably in the region of $20 million.

In the past 40 years, there have been long periods when pyrethrum extract has been in short supply – as it is at present – then shorter periods in surplus. The current shortage means that most usage of pyrethrum is in the USA, as producers have tended to give that country preference with their limited supplies. This roller -coaster availability has encouraged the development and use of synthetic alter - natives, especially outside the USA, but none of these alternatives have all the unique properties of natural pyrethrum.

For the first 130 years of usage, on or near humans, the only safety complaints were due to the appearance of skin allergies. The most vocal complaints were from the British and Allied soldiers in World War II who used pyrethrum ointments based on crude oleoresin extracts to repel mosquitoes whilst fighting in tropical areas. These negative comments led to the refining of the crude extracts. At the same time, the first Freon-based aerosols were being developed which required refined extracts– now described as pale extracts as they are dewaxed and decolourised.

During the next 40 years, pyrethrum was considered as being a very safe product for use on or near humans and animals. From 1930–1955 pyrethrum extract was the anthelminthic of choice for young children in France, and from 1950 it was added in an alcoholic solution to potable water as a simple but effective control of the small water crab Ascellus aquaticus found in water mains in Europe. A very low dose of pyrethrum caused these crabs, which blocked the filters, to lose their grip and then they could be washed away. In Australia from 1963 onwards, many millions of cans of liquid repellents containing up to 0.5%w/w of pyrethrins were applied to human skin without any significant adverse effects being reported. In the USA, up to 1973 household products containing synergised pyrethrum could be labeled as "nontoxic to human and pets" and there has been continuous use of pyrethrum-based formulations to control head lice in children in the Western world since refined extracts became available.

In 1999 the EPA in the USA having examined data from the long-term feeding of extremely large doses of pyrethrum extract to rodents suggested it was "likely to be a human carcinogen by the oral route". In practice the oral daily dosing of up to 5000 ppm during one's lifetime would never take place.

The WHO/FAO Joint Meeting on Pesticide Residues (JMPR) have issued a report on the safety of pyrethrum extract following their recent review (Solecki, 1999). They take a more pragmatic view and because of its importance this is quoted in full:"In a two-year study of toxicity and carcinogenicity in rats and an 18-month study of carcinogenicity in mice,the NOAEL [no observed adverse effect level] was 100 ppm in both species, equal to 14 and 4 mg kg–1body weight per day in mice and rats, respectively . The liver was the main target. A treatment-related effect on the incidence of lung tumours was seen in mice and increased incidences of benign tumours of the skin,liver, and thyroid were observed in rats. The increased incidences of hepatocellular adenomas were associated with persistent induction of cytochrome P450 enzymes and hepatocellular hypertrophy, suggesting that pyrethrins are rodent-specific hepatoproliferative carcinogens. Enzyme induction leading to increased clearance of thyroid hormones would also be consistent with the higher incidence of follicular hyperplasia and follicular adenomas. However, additional studies on the mechanism of formation of the liver and thyroid tumours are required. The Meeting concluded that the increased tumour incidences caused by pyrethrins are threshold phenomena of negligible relevance to the low doses to which humans are exposed."

Table 13.1: The Most Common Organic Pesticide Use with Its Sources

Category	type	Active ingredient	Product name	Targets	Crop
Micro-organism					
Bacteria	insecticide	Bacillus thuringiensis var kurstaki	Dipel DF	caterpillars	vegetables, soft fruit, ornamentals and amenity vegetation
	fungicide	Bacillus subtilis QST713	Serenade ASO	Botrytis spp.	vegetables, soft fruit, herbs and ornamentals
	nematicide	Pasteuria usgae	Pasteuria usgae BL1	sting nematode	turf
Fungi	insecticide	Beauveria bassiana	Naturalis L	whitefly	protected edible and ornamental plant production
	fungicide	Coniothyrium minitans	Contans WG	Sclerotinia spp.	outdoor edible and non-edible crops and protected crops
	herbicide	Chondrostereum purpureum	Chontrol	cut stumps of hardwood trees and shrubs	forestry
	nematicide	Paecilomyces lilacinus	MeloCon WG	plant parasitic nematodes in soil	vegetables, soft fruit, citrus, ornamentals, tobacco and turf
Viruses	insecticide	Cydia pomonella GV	Cyd-X	codling moth	apples and pears
	anti-viral	zucchini yellow mosaic virus, weak strain	Curbit	zucchini yellow mosaic virus	transplanted zucchini and cantaloupes, watermelons, squash
Oomycetes	herbicide	Phytophthora palmivora	DeVine	Morenia orderata	citrus crops
Biochemical	insecticide	azadirachtin	Azatin XL	aphids, scale, thrips, whitefly, leafhoppers, weevils	vegetables, fruits, herbs and ornamental crops
	fungicide	Reynoutria sachalinensis extract	Regalia	powdery mildew, downy mildew, Botrytis, late blight, citrus canker	protected ornamental and edible crops
	herbicide	citronella oil	Barrier H	ragwort	grassland
	nematicide	Quillaja saponaria	Nema-Q	plant parasitic nematodes	vineyards, orchards, field crops, ornamentals and turf

	attractant	citronellol	Biomite	tetranychid mites	apples, cucurbits, grapes, hops, nuts, pears, stone fruit, nursery and ornamental crops
Semio-chemical	attractant	(E,E)-8,10-dodecadien-1-ol	Exosex CM	codling moth	apples and pears

Two types of botanical pesticides can be obtained from seeds of the Indian neem tree, Azadirachta indica (Meliaceae) (Schmutterer 1990, 2002). Neem oil, obtained by cold-pressing seeds, can be effective against soft-bodied insects and mites but is also useful in the management of phytopathogens. Apart from the physical effects of neem oil on pests and fungi, disulfides in the oil likely contribute to the bioactivity of this material (Dimetry et al. 2010; Dimetry 2012). More highly valued than neem oil are medium polarity extracts of the seed residue after removal of the oil, as these extracts contain the complex triterpene azadirachtin. Neem seeds actually contain more than a dozen azadirachtin analogs, but the major form is azadirachtin and the remaining minor analogs likely contribute little to overall efficacy of the extract. Seed extracts include considerable quantities of other triterpenoids, notably salannin, nimbin, and derivatives thereof. The role of these other natural substances has been controversial (El-Sayed 1982– 1983a, b), but most evidence points to azadirachtin as the most important active principle (Isman et al. 1996). Neem seeds typically contain 0.2–0.6 % azadirachtin by weight, so solvent partitions or other chemical processes are required to concentrate this active ingredient to level 10–50 % seen in the technical grade material used to produce their products (Sallena 1989; Schmutterer 1990). Azadirachtin has two profound effects on insects. At the physiological level, azadirachtin blocks the synthesis and release of molting hormones (ecdysteroids) from the prothoracic gland, leading to incomplete ecdysis in immature insects. In adult female insects, a similar mechanism of action leads to sterility. In addition, azadirachtin is a potent antifeedant to many insects. The discovery of neem by western science is attributed to Schmutterer (1990), who observed that swarming desert locusts in Sudan defoliated almost all local flora except for some introduced neem trees (National Research Council 1992). Indeed, azadirachtin was first isolated based on its exceptional antifeedant activity in the desert locust, and this substance remains the most potent locust antifeedant discovered to date.

Unlike pyrethrins, azadirachtin has defied total synthesis to this point. In USA, Neem rapidly became the modern paradigm for development of botanical pesticides (Weinzierl 2000).(source:Botanical Pesticides and Their Mode of Action p.137).

The main advantage of botanicals is that they are easily produced by farmers, cheaper and hazard free in comparison to chemical insecticides (DeVillier et al. 2011). Botanical extracts have defensive compounds which make difficult or impossible for pests to consume the plants. Thus, in many countries efforts are being made to minimize the use of harmful chemical insecticides through the use of indigenous plant products and use of biodegradable products to protect the brinjal plant (Romeis and Kennedy 2013). The plant products include extracts of dried leaves, fruits and seeds.

For over three decades, the Philippine agricultural sector has depended and relied on inorganic fertilizers and pesticides for food production. Because of the lack of an effective and locally available fertilizer and pesticides technologies, the Philippines resorted to the importation of 85% of the total inorganic fertilizer and more than 90% of pesticide requirements. Large amounts of foreign exchange spent on importation have contributed to the stagnant and limited growth of our economy. It is in this regard the bio-pesticide research in the country was undertaken in the late 70s to come up with more cost-efficient and local alternatives to imported fertilizers and pesticides.

With the establishment of the National Institutes of Biotechnology and Applied Microbiology (BIOTECH) at the University of the Philippines Los Baños (UPLB), mycorrhizae research grew very rapidly and gained momentum during the last two decades. "Bio-fertilizer Technology Program", formerly "Nitrogen Fixation and Mycorrhizae Program", became one of the major programs of the institute and this program has been mandated to undertake researches in all aspects of mycorrhizal symbiosis, organic composting and nitrogen fixation. The goal was to look for cheaper and locally available but efficient substitutes or supplements to inorganic fertilizer. Thirty years from its inception, the program has developed several microbial inoculants and bio-fertilizers for agricultural crops, ornamentals and forest trees, which could increase yield/growth of targeted crops, significantly reduce or eliminate chemical fertilizer requirements, and are environmentally sound.

On the other hand, research on the use of plants as bio-pesticides in the Philippines had not gained that momentum as in bio-fertilizers due to inadequate financial support. Most of the bio-pesticide researches in the Philippines have been conducted at the International Rice Research Institute (IRRI) and at UPLB (Departments of Entomology and Plant Pathology). Other research institutions conducted preliminary screening of some crude extracts but results are mostly unpublished. The plant parts being bioassayed are leaves, flowers, stem, vine, bark, fruit peelings, seeds and bulbs. Most of the studies are very preliminary, with crude extracts assayed on a few test organisms.Figure 13.1 shows the effectivity of the leaves extract in eradicating pest using the laboratory experimental procedure using the toxicity level in a brineshrimp bioassay. This laboratory test was further applied by spraying it with the experimental pests such as mosquitoes, coackroaches, flies, and ants. It was found out highly effective in killing this insects at a variable concentrations.It was found out the effectivity of the plant extract is due to the presence of the bioactive components which was tested using the phytochemical screening (table 13.2) of the six bioactive components of alkaloids, anthraquinones, flavonoids,tannins, saponins and steroids.The most important bioactive compound present which can attribute in the killing of pests are the presence of saponins and tannins which could be seen abundantly present in the crude extract of this plant sources.

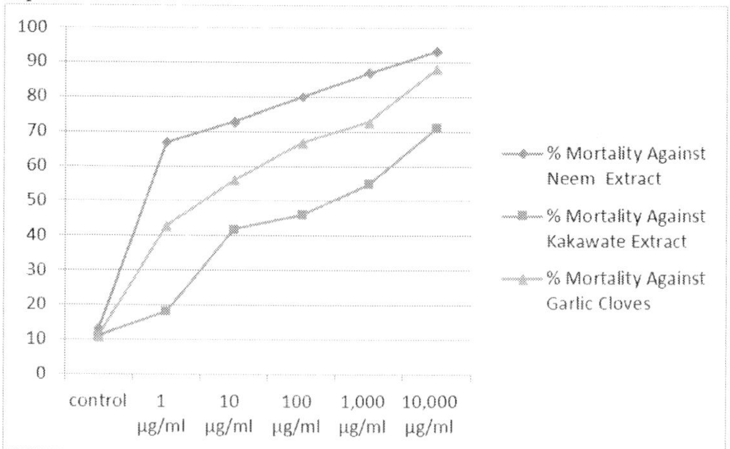

Figure 13.1: The mortality rate of in brineshrimp bioassay in application of the crude extract of plant sources.

Figure 13.1 shows the comparison of the mortality rate of the three plant extracts. The figure farther shows that the mortality rate of neem leaves extract (in blue) is more highly effective followed by the application of garlic cloves extract (in green) and finally the application of kakawate leaves(extract (in red). This could shows that the effectivity as pesticide of the plant extracts are neem leaves> garlic cloves> kakawate leaves extract.

Table 13.2: The Phytochemical Screening of the 3 Sample Plant Extract Tested for the Bioactive Compounds

Plant Sample	**Test for Alkaloids**		**Test for Tannins**		**Test for Saponins**	**Test for Flavonoids**	**Test for Steroids**	**Test for Anthraquinones**
	Mayers Reagent	Dragendorffs Reagent	Gelatin Test	Ferric Chloride Test	Froth Test	Bate & Smith Method	Keller kiliani test	Borntrager's Test
Neem Leaves Extract	+++ w/ppt.	+++ w/ppt.	+++ Brownish color(condensed tannins)	+++ Brownish color(condensed tannins)	+++ More than 30 cm froth	+++ Highly present	Brown ring +++ Highly present	++ Moderately present
Kakawate Leaves *(Gliricidia sepium)* **Extract**	+ w/ slight turbic color	++ w/ moderate turbid color	++ Green black hyrolyzable tannins	+ Slight greenish black for hydrolyzable tannis	++ More than 20 cm froth	- absent	++ Presence of prominent brown ring	– absent
Garlic Cloves Extract	+++ w/ppt.	+++ w/ppt.	+++ Brownish color(condensed tannins)	+++ Brownish color(condensed tannins)	+++ More than 30 cm froth	+ Slightly present	– absent	++ Moderately present

Figure 13.2: The pictures of the different plant extract of Kakawate leaves, Garlic cloves and Neem leaves as sources of organic pesticide

EFFECTIVITY OF THE BOTANICAL PLANT AGAINST MOLLUSCICIDE

Golden Apple Snail

The golden apple snail, popularly known as "golden kuhol" [Pomacea canaliculata Lamarck], was introduced into the Philippines between 1982 and 1984. It came from South America (Brazil and Argentina) via Taiwan. Its high nutritive value as food for human beings and farm animals generated interest among both public and private sectors to propagate the production of this organism. However, a few years after its introduction, the golden apple snail became a major pest of rice.

The golden apple snail has a brachial respiration system that allows it to breathe under water as well as a lung that respirates air. This allows the snail to adapt quickly to changing conditions and expands its ability to search for food, prompting the snail to leave the water when the food supply below the surface is inadequate. Female snails lay up to 500 eggs per week, causing population explosions. Such adaptations, along with high population densities, have made these snails a serious pest in many areas of cultivated rice land in Asia. The snails eat the base of padi seedlings, then consume the aerial leaves and stems, all of which takes an adult snail between three and five minutes.

The only commercial botanical molluscicide registered for use against GAS by the FPA is tannin-glycoside-sterol-flavanoid (Protek, 2010). Protek was developed from natural toxins with specific properties against Pomacea spp., low mammalian toxicity and high biodegradability because of low toxicity (Cyanamid Pro Green, 2011).

Tannins contribute to many aspects of our daily lives. They are responsible for the astringent taste we experience when we partake of wine or unripe fruits, and for the enchanting colors seen in flowers and in autumn leaves.

One of the most satisfactory definition of tannins was given by Horvath (1981): "Any phenolic compound of sufficiently high molecular weight containing sufficient hydroxyls and other suitable groups (i.e. carboxyls) to form effectively strong complexes with protein and other macromolecules under the particular environmental conditions being studied". Tannins can complex with proteins, starch, cellulose and minerals.

In this study, the researchers decided to use the following plants (guava, ipil-ipil and Makahiya plants). These three plants contain tannins which is said to be a good source for a botanical pesticide used in killing golden apple snails. Actually one of the strongest benefits of biopesticides are that they offer a more sustainable solution than synthetic alternatives, or more specifically that they allow chemical pesticides to be deployed where and when they are most needed and most appropriate so that a precious resource is not squandered. Chemical pesticides are withdrawn because of resistance problems or because they are no longer commercially viable, a space is created for biological solutions to occupy. One of the clear benefits of biopesticides is that they can be *combined* with other solutions. They can complement other forms of biological control, e.g., conservation measures such as beetle banks.

They can allow the use of other forms of control with low efficacy, e.g., where there is partial resistance to existing products. They are also relatively cheap to develop and easily gathered hence it is just in the surrounding and commonly found in the rural areas where rice field are abundant. Biopesticide do not have residue problems which are a matter of significant concern for consumers and may deter them from purchasing fresh fruit and vegetables. For retailers' biopesticide offer the possibility of brand security, e.g., they could be applied late on top fruit and would leave no residues.

The golden apple snail (GAS), Pomacea canaliculata (Lamarck), originating from South America, has increased its invasiveness and damage to rice plants in Asia and North America where it was introduced. It is known by an array of common names such as: golden miracle snail, golden snail, jumbo snail, Argentinian apple snail, bisocol and golden kuhol. The term "golden" refers not to their color, but tothe amount of money snail wranglers could make raising them. However, GAS is its frequently used common name. It is listed as one of the World's Worst Invasive Alien Species (IAS) as its invasion ability is based on unique morphological and biological characteristics that support their capacity to survive in adverse environmental conditions and still reproduce fast.

GAS has become a major pest of rice in all the rice -growing countries where it was either intentionally or accidentally introduced (Joshi et al., 2003). The most recent serious invasions are reported from Republic of Dominica, Papua New Guinea and South Korea. GAS continues to be a problem in Japan, Thailand, Vietnam, Cambodia and Malaysia. In Asia, distribution continues to expand westward. Large rice -growing regions of India, Bangladesh, China and Australia are the probable next targets of GAS invasions. From the information gathered in the given countries, it could be generalizedthat: (1) four years after its introduction in a country, it attains the pest status; (2) local establishment is a key feature of invasion; (3) establishment is stochastic; (4) once established, it will typically persist;and (5) invasion-resistance increases during assembly Invasive Alien Species (IAS) as its invasion ability is based on unique morphological and biological characteristics that support their capacity to survive in adverse environmental conditions and still reproduce fast. GAS has become a major pest of rice in all the rice-growing countries where it was either intentionally or accidentally introduced (Joshi et al.,

2003).

The most recent serious invasions are reported from Republic of Dominica, Papua New Guineaand South Korea. GAS continues to be a problem in Japan, Thailand, Vietnam, Cambodia and Malaysia. In Asia, distribution continues to expand westward. Large rice-growing regions of India, Bangladesh, China and Australia are the probable next targets of GAS invasions. From the information gathered in the given countries, it could be generalized that: (1) four years after its introduction in a country, it attains the pest status; (2) local establishment is a key feature of invasion; (3) establishment is stochastic ; (4) once established, it will typically persist; and (5) invasion-resistance increases during assembly.

For health and environment protection, organic farming is becoming popular and demand for organically produced farm products is increasing around the globe. Organic crop producer however are confronted by the problem of controlling usual pest in the crops which tempt them to use commercial pesticides. There is therefore a need to provide farmers with alternative materials that may not be as harmful as synthetic chemical pesticides. In the Philippines, one of the serious problems of organic rice producers is the proliferation of golden apple snail (Pomacea canaliculata) invading almost all rice fields in the lowlands including rice terraces in the uplands.

Another serious problem of organic rice producers particularly in the Ifugao Rice Terraces inscribed by UNESCO as a World Heritage Site in 1995, is the giant earthworm (Pheretima elongata) infestation. The exceptionally large (finger-sized) and long (1-1.6 feet) giant earthworms borrow into the rice terraces soil in its walls and paddy dikes causing water seepage and collapse of the terraces. The giant earthworms are destructive not only significantly reducing organic rice production yield but also seriously affecting the tourism industry of the province due to the damage of the rice terraces. Infestations by Pomacea canaliculata and Pheretima sp. are causing significant damage to the production of organic "Tinawon" Heirloom Rice that is in demand locally and internationally. To maintain the organic quality of heirloom rice, farmers need to hand-pick golden apple snails in their rice fields but is very laborious and time consuming. For Pheretima, it is very hard to control because they borrow inside the soil. While application of synthetic chemical pesticides can easily control these two serious pests, it is not being practiced and its use is aggressively discourages mainly because it is against the practices of organic farming. Besides, chemical pesticides are costly, create toxic risks to human and not-target organisms, and cause environmental problems such as soil and water pollution.

The promotion of the practice of organic farming in the Philippines through Organic Agriculture Act of 2010 suggests that it is timely to use not only organic fertilizers but also organic pesticides. Since it is organic, it does not pose any hazard to farmers and the environment. In the Philippines, some plant species are known to have molluscicidal properties like Sandorium vidalii ((Ngaloy and Andrada 2005), and seeds of Sandoricum Koetjape (Taguiling and Buyucan 2008). Plants with vermicidal property include Harpulia Arborea (Taguiling and Ngabit, 2010) and Parkia sp.

Other Organic molluscicides include the use of Tobacco dust as an organic, readily degradable, and environment-friendly molluscicide (Borja 2012; Tangkoonboribun and Sassanarakkit 2009) and an Organic Molluscicide Tea Seed Meal (TSP15) developed by Hangzhou Choisun Tea Sci-tech Co., LTD in China. Aquatic molluscicide in other countries include the use of Endod (Phytolacca dodecandra), which produces a saponin (Mott 1987, Lambert 1991, Ndamba , J. & S.K. Chandiwana 1990); extract of the seeds of Millettia thonningii which is known to be molluscicidal in West Africa; and dried leaf material of a Mexican plant, Agave attenuata, which was found to be highly toxic to the target snail, Bulinus africanus, intermediate host of Schistosoma haematobium in South Africa (Brackenbury & Appleton 1997); and, Detarium microcarpum, Ximenia americana and Polygonum limbatum, known to have molluscicidal activity against aquatic snails particularly in Nigeria (Kela et al. 1995). Most of these plants appear to work as aquatic molluscicides affecting surface tension at the gills of aquatic snails and leading to toxic haemolytic effects. Park (2012) working on medicinal plant extract revealed that extracts from the root or bark of Picrasma quassioides is used as a vermicide or the like. Rawi et al. (2011) investigated the effect of some natural plants (Agave filifera whole, Ammi majus flowers and leaves and Canna indica flowers and leaves) as molluscicidal agents against Biomphalaria alexandrina snails.

The forgoing demonstrated that indeed there are plants that can be used as organic pesticides. There is however a need to evaluate the effectiveness of combination of plant extracts that can be used as organic pesticide to control a wider range of pests. The present study used a combination of some plant extracts to control Pomacea canaliculata and Pheretima sp. This paper presents the result of evaluation of the toxicities of combinations of crude extracts of Sandorium vidalii, Harpulia Arborea and Parkia sp. on golden kuhol (Pomacea canaliculata) and giant earthworm (Pheretima sp.) in support to organic rice farming. This effectivity could be explained due to the presence of bioactive compounds in the plant extract especially in the presence of saponins and tannins.

BIOACTIVE COMPOUNDS

What are Saponins?

Saponins are glucosides with foaming characteristics. Saponins consist of a polycyclic aglycones attached to one or more sugar side chains. The aglycone part, which is also called sapogenin, is either steroid (C27) or a triterpene (C30). The foaming ability of saponins is caused by the combination of a hydrophobic (fat-soluble) sapogenin and a hydrophilic (water-soluble) sugar part. Saponins have a bitter taste. Some saponins are toxic and are known as sapotoxinSaponins are glucosides with foaming characteristics. Saponins consist of a polycyclic aglycones attached to one or more sugar side chains. The aglycone part, which is also called sapogenin, is either steroid (C27) or a triterpene (C30). The foaming ability of saponins is caused by the combination of a hydrophobic (fat-soluble) sapogenin and a hydrophilic (water-soluble) sugar part. Saponins have a bitter taste. Some saponins are toxic and are known as sapotoxin.

Saponins are a class of secondary plant metabolites with diverse biological properties. They occur in a great number of plant species (mainly Angiosperms), both wild plants and cultivated crops. Triterpenoid saponins are mostly found in dicotyledonous species, while many of the major steroidal saponins are synthesized by monocots, such as members of the Liliaceae, Dioscoraceae and Agavaceae families.. Saponins possess clear insecticidal activities: they exert a strong and rapid-working action against a broad range of pest insects that is different from neurotoxicity. The most observed effects are increased mortality, lowered food intake, weight reduction, retardation in development and decreased reproduction. Accordingly, saponins exert a repellent/deterrent activity, bear digestive problems, provoke insect moulting defects or cause cellular toxicity effects. As a con-sequence these interesting plant components open new strategies to protect crops in modern agriculture and horticulture with integrated pest management (IPM) programs against pest insects.

What are Tannins?

Tannin compounds are derivatives of tannic acid and are characterized by cyclic structures coupled with hydroxyphenols. The tannins are divided into two main groups, condensed tannins and hydrolyzable tannins. The condensed tannins have been identified in many plants. Although the role of these tannins is not clear, it has been postulated that they act to protect the plant from insect, fungal, etc.,.Condensed tannins are characterized by the presence of a large number of phenolic groups, and, thus, will tightly bind to amides, such as those present in proteins. Any such attachment to a protein will induce a conformational stress which can eventually denature or precipitate it, or block its functional groups. Thus, plan tannins can act to inhibit the actions of any protein-based toxin produced by a bacterium such as BT. Also, ingestion of tannin-containing plants by insects will cause a high concentration of tannin in the insect's gut; such tannins.

What are Alkaloids?

Alkaloids are basic nitrogenated compounds, comprising the largest class of secondary plant metabolites. There are over 5, 500 compounds of alkaloids. Alkaloids have been found to exhibit dramatic physiological effects, making them valuable in the field of medicine. They are widely used as cancer chemotherapeutic agents (Chabner and Horwit, 1990) and also interfere with cell division as the findings of Chewonarin, *et al.* (1999) that isolated an alkaloid and demonstrated its ability to prevent mutagenesis.

What are Flavonoids?

Flavonoids are organic pigments occurring in plants which give plants a red, violet or blue color. Flavonoids inhibit the growth of bacteria and viruses, protect the cells against the damages of free radicals, protect against cancers and heart attacks, have a repressive effect against inflammations and they influence blood coagulation (novafee.com/nutrition/plant-secondary-metabolites.htm/Accessesd200914March).

What are Anthraquinones?

Is polycyclic aromatic hydrocarbon derived from anthracene or phthalic anhydride. These were the main active constituents in herbs often used to relieve constipation. They have an irritant or stimulating laxative effect on the large intestine. Anthraquinones act directly on the intestinal mucosa, influencing several pharmacological targets, and their laxative effect was largely due to increased peristalsis of the colon, reducing transit time and consequently the reabsorption of water from the colon making the stool more liquid and easing bowel movements.

What are Steroids?

Plant steroids are derived from sterols and comprise steroid saponins, steroid alkaloids, pregnanes , androstanes, estranges, ecdysteroids, with anolides and cardiac glycosides. Steroids are widely used as drugs constitute anti-inflammatory, contraceptive, and anti-cancer agents (Wink, 2003). Plant sterols and stanols (type of steroids), are found naturally in many plant foods including vegetable oils, nuts, fruits and vegetables. They have chemical structures that are similar to cholesterol, and it seems that they affect the absorption of dietary and biliary cholesterol in the intestinal lumen, replaying it in the mixed micelles (Suaer *et al.*, 2010).

STUDIES ON THE APPLICATION OF PLANT EXTRACT TO SNAIL

Experimental application of the plant extract samples of Ipil-ipil leaves(*Leucaena glauca L.*) , guava leaves(*Psidium guajava*) and

Makahiya leaves (*Mimosa pudica*) (figure 13.4) which was tested in the laboratory for the effectivity in application for the golden apple snail mulluscicidal activity. It was found out that plant extract from botanical plant sources can be effective substitute in golden apple snail infestation which becomes a problem of the farmers. This golden apple snail has the tendency of eating up the rice plant at the mature stage by attaching itself in the rice plant as the source of their food. Eating it up the stem, leaves and most of all the stalk that contains the grains of the rice. It also lays eggs in the leaves of the rice and multiply up to hundred of another snails which in turn totally kills the plant if not been detected as shown in figure 13.3.

Figure 13.3: Picture of Golden Apple Snail infestation in a rice field.

Figure 13.4: Pictures of different plant extract use in the experimental application for the mortality rate of Golden Apple Snail

The application of the plant extract against the golden apple snail showed a higher percentage of mortality (figure 13.5 to figure 13.7) this is due to the presence of the bioactive compounds especially tannins due to the large number of phenolic groups which will tightly bind to amides such as those present in proteins that inhibits the action of insectsto destroy the plant and saponins exert a repellent/ deterrent activity, bear digestive problems, provoke insect moulting defects or cause cellular toxicity effects on the insects that harms the plants.

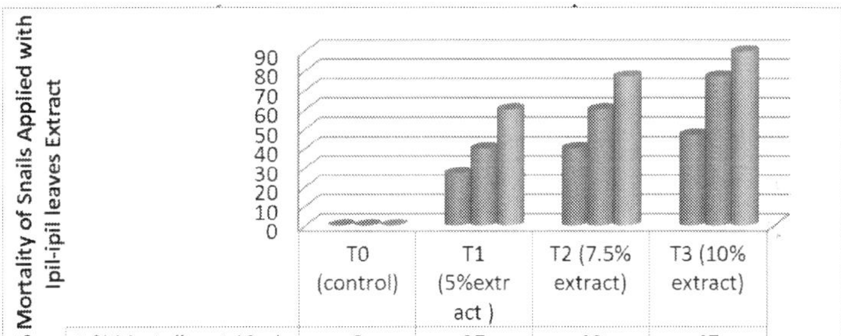

	T0 (control)	T1 (5%extr act)	T2 (7.5% extract)	T3 (10% extract)
% Mortality at 10min	0	27	40	47
% Mortality at 20min	0	40	60	77
% Mortality at 30min	0	60	77	90

Figure 13.5: Graphical Representation of the Mortality rate of Plant extract Ipil-ipil leaves (*Leucaena glauca L*)Against Golden Apple Snail

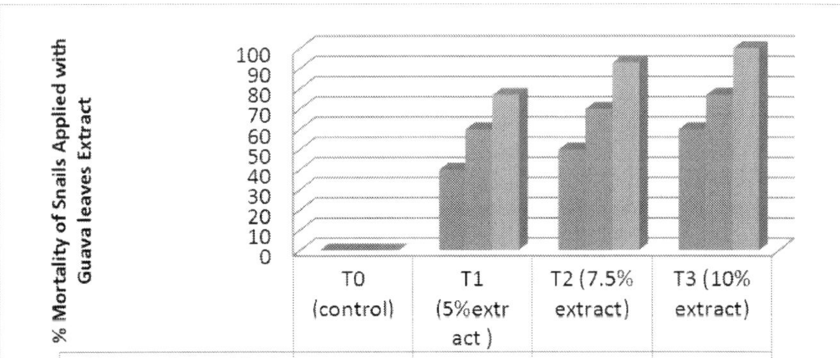

	T0 (control)	T1 (5%extr act)	T2 (7.5% extract)	T3 (10% extract)
% Mortality at 10min	0	40	50	60
% Mortality at 20min	0	60	70	77
% Mortality at 30min	0	77	93	100

Figure 13.6: Graphical Representation of the Mortality rate of Plant extract Guava Leaves (*Psidium guajava*) Against Golden Apple Snail

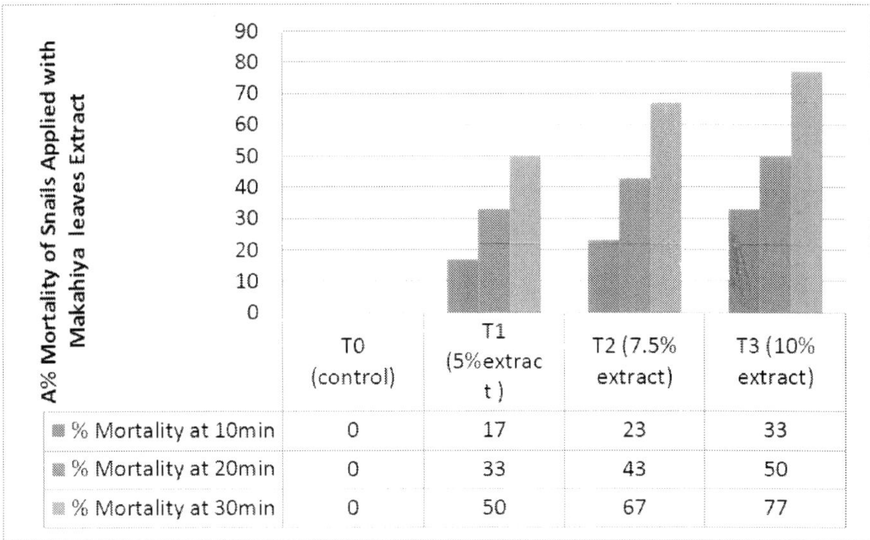

Figure 13.7: Graphical Representation of the Mortality rate of Plant extract Makahiya Leaves (*Mimosa pudica*) Against Golden Apple Snail

Table 13.3: The Phytochemical Screening of Botanical Plants of Ipil-ipil, Guava and Makahiya Leaves Extract

Plant Sample	Test for Tannins		Test for Saponins
	Gelatin Test	Ferric Chloride Test	Froth Test
Ipil-ipil leaves Extract	++	Brownish Brown ++	+++
Guava Leaves Extract	+++	Brownish Brown +++	+++
Makahiya Leaves Extract	+	+	++

Another example of the experimental process in the application of botanical plant comprising of five highly potential plants, Annona squamosa seed, Nerium indicum Leaves, Stemona tuberose root, Cyperus rotundus corm and Derris elliptica root was assessed to Pomacea canaliculata. D. elliptica root and C. rotundus corm extracts with its toxicity activity as shown in figure 13.8. The phytochemical analysis (Table 13.4) showed that the combined extracts tested as organic molluscicide/vermicide contains compounds such as alkaloids, deoxysugars, saponins and unsaturated sterols and triterpenes . Compounds with high potential of having toxic effect to golden kuhol and giant earthworm include alkaloids, saponins and unsaturated sterols and triterpenes.

Many alkaloids are toxic to other organisms especially plants that contain tropane alkaloid or isoquinoline alkaloid. Glycoside is toxic when the compound contains cyanide group. Saponin content of bark of some plants is highly toxic and can be applied as agricultural spray. Saponins also exhibit hemolytic properties, which act as poison, show cytotoxic or pesticidal activity, and have a variety of medicinal applications. Together with this species with stronger properties would consequently be more active than combining species together with weaker properties. This may explain why extracts from combination of Harpulia and Parkia appeared to be the least effective . According to Andrada and Ngaloy (2005) Malasantol contains considerable levels of tannin, saponins and alkaloids which may be toxic to have caused mortality of golden apple snail. Between dosages, higher concentration resulted to shorter time of mortality of golden kuhol and in agreement with the findings of El-Din et al. (2011) stating that the percentage of dead snails (Biomphalaria alexandrina), at all stages increased with increasing concentration of extract of Dalbergia sissoo (family Leguminoseae). Crebassa et al. (2012) revealed that the active ingredient of the plant extract from Gleditsia species was Saponin.

Table 13.4: Results of phytochemical analysis of organic pesticide

Test for the Phytocompound Present	Results	Bioactive Compound Present
1) Alkaloids (Mayer's test)	+	Alakaloids
2) Steroids	+	Deoxysugars
2.1) Keller-Killani	+	Unsaturated sterols
2.2) Lieberman-Buchard	-	Unsaturated lactones
2.3) Kedde test	-	
3) Anthraquinones	-	Anthraquinones
4) Flavonoids	-	Leucoanthrocyanins and cyanindins
5) Saponins		Saponins

5.1) Froth tests	+	Unsatorated sterols & triterpenes
5.2) Liebermann-Buchard	+	
6) Tannins & polyphenols		Tannins
6.1) Gelatin test	-	Polyphenolic compounds
6.2) Ferric chloride test	-	
7)Cyanogenic glycoside	+	Cyanogenic glycoside

The molluscicidal activity of crude extracts from five highly potential plants, Annona squamosa seed, Nerium indicum Leaves, Stemona tuberose root, Cyperus rotundus corm and Derris elliptica root was assessed to Pomacea canaliculata. D. elliptica root and C. rotundus corm extracts showed the highest toxicity against 3-month old snails which have LC_{50} as 23.68 +/- 2.96 mg/l and 133.20 +/- 7.94 mg/l, respectively. The C. rotundus corm extracts were chosen for detoxification enzyme in vivo assay which shows esterase and glutathione S-transferase activity in stomach, intestinal tracts and digestive glands of survival treated P. canaliculata were inhibited.

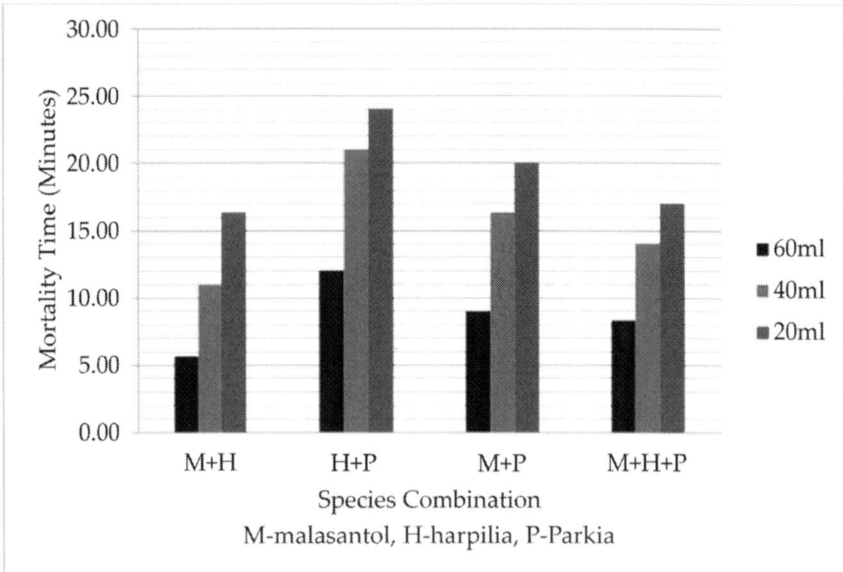

Figure 13.8: Molluscicidal activity of combined crude extracts of species against Pheretima sp. worms.

Golden apple snails are harmful mollusks found in the farm especially rice fields, which can be feed with different plants such as algae, water hyacinth, rice seedling, grown rice or mature rice or any succulent leaves and stem. This kind of snail loves to eat young soft parts of the plants because of its rough tongue that sucks and scrapes the plants stem or leaves. They could also eat any dead animals and usually attack during night time and actively doing its hazard during this time in order to for them to make havoc without knowing it.

Golden apple snails had caused a problem with the problems due to its rapid reproduction. A female snail could lay eggs in the water every early morning or evening which could not be often notice with about 25-500 bright pink eggs. These eggs are attach in the rice stalk or stem in the rice field or in the upper part of the water and within 10 to 15 days these pink eggs will hatch on hundreds of snails which will again add on to the problematic presence of this snails.

Golden apple snails love to stay in the fresh water but it could also live in a little salty water or swampy areas. At a very low temperature of 0 degree Celsius it will die. This type of snail can be found in areas wherein it is a little bit humid with rice paddies and threaten the farmers source of livelihood of the farmers. Countries such Philippines, South America such Brazil and Uruguay which this pests had been originated. This was exported to the different countries in Asia especially Vietnam and Philippines because its main purpose will a good source of protein for farmer whose main staple food is rice.

The exportation of the golden apple snails has become a trouble to Asian countries because instead of becoming source of food, it comes as pests. Thus, this snail has become a common enemy of farmers because they eat their way out to the green fields of farmers and threaten the means of livelihood of farmers in Asian countries likes Korea, Japan, Vietnam and even Philippines.

BIOLOGICAL AND PHYTOCHEMICAL SCREENING OF BOTANICAL PLANT EXTRACT AS PESTICIDE

One of the studies involved in the efficacy of botanical plant extracts as pesticide is the application of capsicum which is widely use alternative in most Asian countries in organic farm. A laboratory experiment was investigated involving three ethanol extracts from three varieties of capsicum species namely; the Chili(*Capsicum annuum*), Bell pepper (*Capsicum frutescence*) and Kulikot (*Capsicum longum)* were collected and assayed to test the bactericidal activity against 2 bacteria; *Escherichia coli* (gram negative) and *Staphylococcus aureus* gram positive).

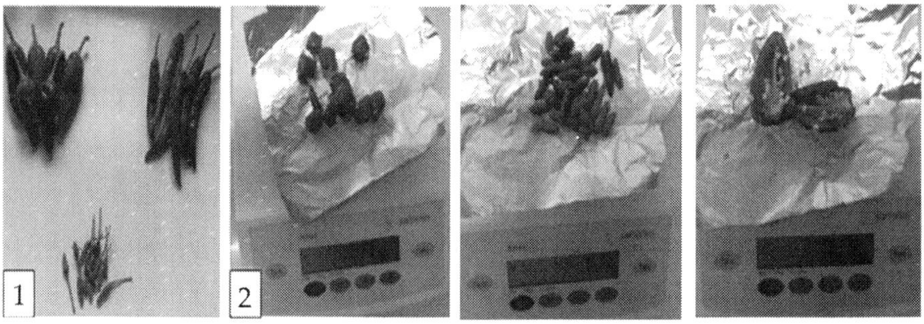

Figure 13.9: Picture of the Three Varieties of Capsicum as Biopesticide

Suspensions were prepared using MHA (Mueller-Hinton agar) agar dilution and nutrient broth solution. The said bacteria were swab into culture media and their possible antibiotic resistance was determined by measuring the zone of inhibition using paper disc-diffusion method. The minimum inhibition concentration (MIC) of the three capsicum species: Chili (*Capsicum annuum*), Bell pepper (*Capsicum frutescence*) and Kulikot (*Capsicum longum)* were 10 g/20 ml of 95% of ethanol. The three extracts were evaluated for bactericidal activity. The extracts of Chili (*Capsicum annuum*), Bell pepper (*Capsicum frutescence*) and Kulikot (*Capsicum longum)* showed excellent bactericidal effect against *Escherichia coli* and *Staphylococcus aureus.*

The study investigated the effectivity of Capsicum using the Phytochemical screening of the three widely known *Capsicum* species fruits considering only to the five bioactive secondary metabolites, which includes the alkaloids, saponins, flavonoids, steroids, anthraquinones, and tannins. This study was delimited only to the fruits of capsicum species and assayed through 95 % ethanol as extracting solvent. *Escherichia coli* and *Staphylococcus aureus* are the test microorganism for the antimicrobial activity test. A Brine Shrimp Bioassay was also employed to test the toxicity level to the 48 hours nauplii. Chrorioallantoic Membrane Assay was carried out to assess whether extracts were pro or anti Angiogenic agaents in the 12 day old duck embryo (Anas platyrynchos).

The bactericidal activity of selected plant spices against the mentioned microorganisms and their effectiveness were assessed by the presence or absence of inhibition zones, the bactericidal experiment with 6 petri dish having 3 replicates of each experimental and control group (table 1.5). The concentration of the fresh plant extract use in this experiment was 20 grams of each capsicum species diluted into 20 ml of pure 95% ethanol. The capsicum sp. are Bell pepper represent as 1, Chili represent as 2 and Kulikot represented as 3 which were the experimental group , the positive control of the experiment was an antibiotic named Cotrimoxazole(+) while the negative control was the 95% ethanol(-). The test microorganisms were the *Escherichia coli* and *Staphylococcus aureus.* The size of the paper disc used is 6 mm in diameter.

Table 13. 5: Dilution series of plant extracts diluted with distilled water in a dose of 1, 000, 000 ppm.

Treatment		**Diluent (mL)**	**Dose of Extract (ứL)**	**Total Vol. in 2 replicate paper discs (mL)**
Kulikot *(Capsicum frutescence)*		1	90	1.09
Bell Pepper *(Capsicum longum)*		1	75	1.075
Chili *(Capsicum annum*	1	75	1.075	

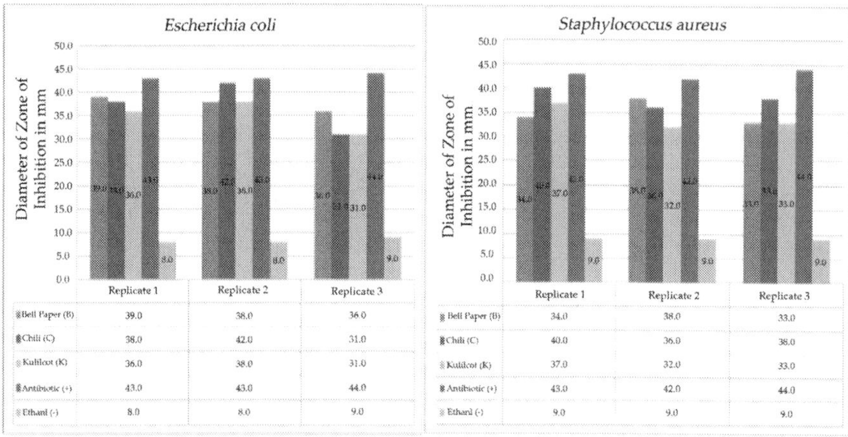

Figure 13.9: Bacterial Activity of Three Varieties of Capsicum Against *E. coli* and *S. Aureus*.

Among the Three *Capsicum* sp. tested, all the three showed antibacterial activity. The results of the bactericidal activity against the tested pathogens were given in the Figure 5 and 6. However, the different *capsicum* sp. differs significantly in their activity against test microorganisms. The zone of inhibition in Bell pepper in *Escherichia coli* was bigger compare to *Staphylococcus aureus.* In Chili and Kulikot, the zone of inhibition was more or less the same. In the positive control (+),*Staphylococcus aureus* zone of inhibition is much bigger than E.C zone of inhibition. In negative control, *Staphylococcus aureus* zone of inhibition is 9 mm. while *E.coli* was 8mm (figuree 13.9 & 13.10).

In general the experimental group had very active antimicrobial effect on the bacteria *Escherichia coli* and *Staphylococcus aureus*. Hence, the negative control ethanol had inactive effect

Table 13.6: Statistical Analysis of the the Bacterial Activity of Three Varieties of Capsicum

One-way ANOVA: Two-Factor With Replication-E,coli

SUMMARY	Count	Sum	Average	Variance
Bell Pepper (B.)	3	113	37.66667	2.333333
Chili (C.)	3	111	37	31
Kulikot (K.)	3	105	35	13
Antibiotic (+)	3	98	32.66667	4.333333
Ethanol (-)	3	25	8.333333	0.333333
Replicate 1	5	156	31.2	170.7
Replicate 2	5	158	31.6	186.8
Replicate 3	5	138	27.6	112.8

ANOVA

Source of Variation	SS	df	MS	F	P-value	F crit
Rows	1827.73333	4	456.9333	68.36908	3.19E-06	3.837853
Colimns	48.5333333	2	24.26667	3.630923	0.075497	4.45897
Error	53.4666667	8	6.683333			
Total	1929.73333	14				

One-way ANOVA: Two-Factor With Replication-S, aureus

SUMMARY	Count	Sum	Average	Variance
Bell Pepper (B.)	3	105	35	7
Chili (C.)	3	114	38	4
Kulikot (K.)	3	102	34	7
Antibiotic (+)	3	111	37	37
Ethanol (-)	3	27	9	0
Replicate 1	5	161	32.2	175.7
Replicate 2	5	155	31	160
Replicate 3	5	143	28.6	128.3

ANOVA

Source of Variation	SS	df	MS	F	P-value	F crit
Rows	1779.6	4	444.9	46.58639	1.39E-05	3.837853
Colimns	33.6	2	16.8	1.759162	0.232703	4.45897
Error	76.4	8	9.55			
Total	1889.6	14				

Statistical Analysis was performed using one-way Analysis of Variance (ANOVA): two factor with replication. Tests were used for experiments with comparisons of replicates that are provided by the *P values* which is *(<0.05) level of significant* in the table 13.6.

In analyzing the result, <10 mm zone of inhibition means it has inactive antimicrobial effect, 10-13 mm (partially active), 14-19mm (active) and >19 mm (very active). The bactericidal activity of the *Capsicum sp.* extracts and theirpotency was assessed by the presence or absence of inhibition zone as given in Figure 5 and 6. The most susceptible organisms were *Escherichia coli* and *Staphylococcus aureus*, which were sensitive to 3 *capsicum* sp. extract (Chili, Bell pepper and Kulikot). *Escherichia* Coli and *Staphylococcus aureus* were being sensitive to 3 *capsicum*sp. extracts. The maximum inhibitions were observed with the extract of Bell pepper with *Escherichia coli,* the Bell pepper inhibition zone were 39mm for replicate 1, 38mm for replicate 2 and 36mm for replicate 3 with an average of 37.66mm, respectively, for Chili inhibition zone were 38mm for replicate 1, 42mm for replicate 2 and 38mm for replicate 3 with an average of 39.33mm and lastly the Kulikot inhibition zone were 36mm for replicate 1, 31mm for replicate 2 and 31mm for replicate 3 with an average of 32.66mm. While the *Staphylococcus aureus* inhibition zone against produced by the extract of 3 *capsicum sp*, the Bell pepper inhibition zone were 34mm for replicate 1, 38mm for replicate 2 and 33mm for replicate 3 with an average of 35mm, respectively, for Chili inhibition zone were 40mm for replicate 1, 36mm for replicate 2 and 38mm for replicate 3 with an average of 38mm and lastly the kulikot inhibition zone were 37mm for replicate 1, 32mm for replicate 2 and 33mm for replicate 3 with an average of 35.33mm.

Table 13.7: Phytochemical Screening Result for the Three Varieties of Capsicum

Plant Sample	Test for Alkaloids		Test for Tannins		Test for Saponins	Test for Flavonoids	Test for Steroids		Test for Anthraquinones
	Mayers Reagent	Dragendorffs Reagent	Gelatin Test	Ferric Chloride Test	Froth Test	Bate & Smith Method	Keller kiliani test	Buchard Test	Borntrager's Test
Kulikot Fruit (*Capsicum Frutescens*)	++ w/moderate turbidity	+++ w/ppt.	+++ Brown ring	+++ Green black hyrolyzable tannins	+++ More than 30 cm froth	+ present	Brown ring +++ Highly present	+ present	+ present
Bell Pepper Fruit (*Capsicum langum*)	++ w/ moderate turbid color	+++ w/ ppt.	+++ Brown ring	+++ Brownish color (condensed tannins)	++ More than 20 cm froth	++ Moderately present	+++ Presence of prominent brown ring	+ present	+ present
Chili Fruits (*Capsicum annum*)	++ w/ moderate turbidity.	+++ w/ppt.	+++ Brown ring	+++ Green black hyrolyzable tannins	+ 10 cm froth	+ Slightly present	+++ Presence of prominent brown ring	+ present	+ present

Legend: +++ (Highly present); ++ (Moderately present); + (present) and - (absent)

The above table shows the presence of the bioactive compounds in the three varieties of capsicum. It was found out that all of the six bioactive components of alkaloids (abundantly present), tannins (abundantly present), saponins (abundantly present), flavonoids (present), steroids(present) and anthraquinones (present) in the tested three varieties. This farther shows that capsicum has a pesticidal components which could be attested from the presence of the phytocompounds in the plant extract samples and a good insecticidal, herbicidal and molluscicidal effect.

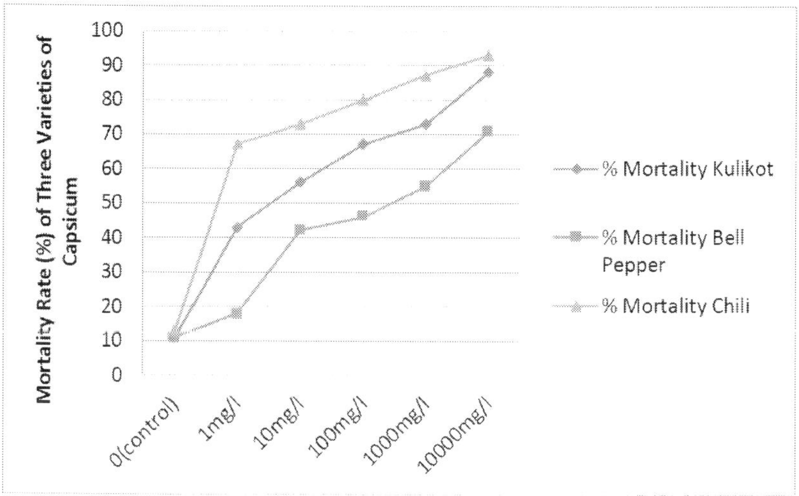

Figure 13.11: Graphical Presentation of Mortality Rate Comparison Between the Varieties of Capsicum

Figure 13.11 above represents the graph in comparison between the three varieties of capsicum and its showed that chili has a higher mortality rate (in green) followed by the mortality rate of kulikot (in blue) and the last is the mortality rate of bell pepper (in red) against nauplii larvae in application to brine shrimp bioassay. It can be deemed further that its effectivity rate of the capsicum varieties are chili>kulikot> bell pepper as shown in figure.

Then primary results showed that the 3 species had very active antimicrobial effect on the bacteria *Escherichia coli* and *Staphylococcus aureus* since the zone of inhibition was more than 19 mm. Chili has the very high antimicrobial effect. Next was the bell pepper and lastly is the Kulikot. But in general, the species have very active bactericidal effect.

Natural extract from selected *Capsicum* sp. have played significant role in the search of biopesticide which is environmentally friendly. The results of the present study were quite encouraging as almost all spices exhibited bactericidal activity against most of the pathogens, but the bactericidal activity varies widely, depending on the type of spices, test medium and microorganism. Bactericidal activity may be dependent on the cell wall composition of bacteria. The use *Capsicum* sp. extracts could be beneficial for farmers hence it could be planted with less cost compared with synthetic pesticide.Further, this kind of biopesticide has no residual effect either in soil and product and in effect this will have health benefits since their is no chemical residue that can be obtain by the consuming public.

REFERENCES

1. Ames, B. N (1993). Phantom Risk, Scientific Inference and the Law . Eds. K. R. Foster , D. E. Bernstein and P .W. Huber. MIT Press, London, UK and Cambridge,USA.
2. Biological Sciences› Miscellaneous Papers. Volume: 68, Issue: 2, Pages: 201-213
3. Borja AO. 2012. Tobacco dust as organic molluscicide. The Philippine Star Jan. 22, 2012.
4. Botanical
5. Brackenbury TD, Appleton CC. 1997. A comprehensive evaluation of Agave attenuata, a candidate plant molluscicide in South Africa.
6. Crebassa X, Caruel H, Mercier F. 2012. Use of Gleditsia Saponins for Controlling Mollusc Pests. US Patent application number: 20120321688:Available online at http://www.faqs.org/patents/app/20120321688#ixzz2VOmwvAfV
7. Crombie, L. (1995). Chemistry of Pyrethrins. Chapter 8 in 'Chemistry of Pyrethrins in the Book "PyrethrumFlowers – Production, Chemistry, Toxicology and Uses'. Ed. J. E. Casida and G. B. Quistad.
8. El-Din AS, El-Sayed K, Mahmoud M. 2011. Effect of ethanolic extract of Dalbergia sissoo plant parts on Biomphalaria alexandrina snail, the intermediate host of Schistosoma mansoni.Medical Malacology Laboratory.
9. FAO/WHO Evaluations 1999. Part II. Toxicological,Pyrethrum. WHO/PCS/00.4, Joint FAO/WHO Evaluations. Part II. 273–291.

10. Journal of Evolutionary Biology Research Vol. 3(7), pp. 95-100. Available online at http://www. academicjournals. org/jebr,
11. Kela SL, Ogunsusi RA, Ogbogu VC, Nwude N. 1989. Screening of some Nigerian plants for molluscicidal activity.
12. Lambert JD, Temmink JH, Marquis J, Parkhurst RM, Lugt CB, Lemmich E, W olde-Yohannes L, De SavignyD. 1991. Endod: Safety Evaluation of a plant Molluscicide.RegulatoryToxicology and
13. Mott KE.1987. Plant Molluscicide. UNDP/W orld Bank/W HO Special Programme for Research and Training in Tropical Disease. John W iley and Sons Ltd.
14. Ndamba J, Chandiwana SK, Makaza N. 1989. The use of Phytolaccadodecandraberries in the control of trematode-transmitting snails in Zimbabwe. Ata Tropica.. 46:303–309
15. Ngaloy IT, Andrada MG. 2005. Effect of Pulverized Dried Bakuwog Fruit on Golden Snail. Ecosystems Research Digest. 15 No. 2. ERDS-DENR CAR, Baguio City.
16. Oxford University Press (1995).Glynne Jones, G. D. (1999). Piperonyl butoxide.Pesticide Outlook 10, 117–120.
17. Park JS. 2012. Medicinal plants extract using processing of herbal medicine. US patent application number 20120039928
18. Pharmacology. 14:189–201
19. pheretima elongata) destroying the Ifugao rice terraces. The Upland Farm Journal 20:26-35.
20. Rawi SM, Al-Hazmi M, Seif Al Nassr F. 2011. Comparative Study of the Molluscicidal Activity of Some Plant Extracts on the Snail Vector ofSchistosoma mansoni, Biomphalaria alexandrina. International Journal of Zoological Research, 7: 169-189
21. Rev Elev Med Vet Pays Trop.1989;42(2):195-202.
22. Solecki, R. (1999). Pesticide Residues in Food. Joint
23. *Sources: University of Arkansas, the Ohio State University, U.S. EPA.*
24. Staudinger , H.; Ruzicka, L. (1924). Insektent?tendestoffe. I-VI and VIII-X. Helvetica Chimica Acta7,177–259, 377–390, 406–458.

25. Taguiling NK, Buyucan M, Baliwang E. 2013. Pofile, Phenology and Propagation trials of some Indigenous Pesticidal Plant Species. The Upland Farm Journal 21:1-18.
26. Taguiling NK, Buyuccan M. 2009. Effect of Milled Santol (Sandoricum Koetjape) Fruit on Golden Apple Snail (Pomacea Canaliculata Lam.). The Upland Farm Journal 18:88-97.
27. Taguiling NK, Ngabit D. 2010. Harpulia arborea: Effective vermicide against giant earthworm (
28. Tangkoonboribun R, Sassanarakkit S. 2009. Molluscicide from Tobacco W aste. Journal of agricultural science. 1:76-81
29. Tattersfield, F., (1931). Pyrethrum Flowers – A Quanti-tative study of their Development. Annals of Applied Biology ,18,602–635.
30. Ahmad, I. and Z.A. Beg, 2001. Antimicrobial and phytochemical studies on45 Indian medicinal plants against multidrug resistant humanpathogens.*J. Ethnopharmacol*., 74: 113–123.
31. Ates, D.A. and O.T. Erdogrul, 2003.Antimicrobial activities of variousmedicinal and commercial plant extracts.*Turkish J. Biol*., 27: 157–162.
32. Babita, P., D. Hari., S. Jin., G. Soon., W. Hyun., 2008.
33. Antibacterial potential of Antartic lichens against Human pathogenic Gram-positive bacteria. Phytother. Res. 22: 1269-1271.
34. Buwa, L.V. and J. Van Staden, 2006. Antibacterial and antifungal activityof traditional medicinal plants used against venereal diseases inSouth Africa. *J. Ethnopharmacol.,* 103: 139–142
35. Baur AW, Kirby WMM, Sherris JC, Truck M (1966) Antibioticsusceptibility testing by standardized single disc method . Am. J. Clin.Pathol. 45: 493-496.
36. Costa C., E. Hiruma-Lima *et al.,* 2008.Antimicrobial activity of some medicinal plants of the Cerrado, brazil.
37. Edeoga, H., D. Okwu, B. Mbaebie. 2005. Phytochemical constituents of some Nigerian medicinal plants. African Journal of Biotechnology. Vol. 4(7) pp. 685-688.
38. Elonima EI, Ahmed SA, Mekkawi AG, Mossa JS (1983). Theantimicrobial activity of garlic and onion extracts. Pharmazie.38: 747-748.

39. Hinou, JB., C. Harvala& E. Hinuo.1989 Antimicrobial activity screening of 32 common constituents of essential oils.Pharmazie 44:302-303.

40. Nakatani, N. (1994). Antioxidative and antimicrobial constituentsof herbs and spices, pp. 251–272. In: CHARALAMBOUS,G. (ed.) Spices, Herbs, and Edible Fungi. Elsevier, Amsterdam

INDEX